OCT 0 6 1989

S0-FKL-729

XX

WORLD BANK TECHNICAL PAPER NUMBER 94

Technological and Institutional Innovation in Irrigation

Guy Le Moigne, Shawki Barghouti, and Herve Plusquellec, editors

Proceedings of a workshop held at the World Bank, April 5–7, 1988

The World Bank
Washington, D.C.

Technical Papers are not formal publications of the World Bank, and are circulated to encourage discussion and comment and to communicate the results of the Bank's work quickly to the development community; citation and the use of these papers should take account of their provisional character. The findings, interpretations, and conclusions expressed in this paper are entirely those of the author(s) and should not be attributed in any manner to the World Bank, to its affiliated organizations, or to members of its Board of Executive Directors or the countries they represent. Any maps that accompany the text have been prepared solely for the convenience of readers; the designations and presentation of material in them do not imply the expression of any opinion whatsoever on the part of the World Bank, its affiliates, or its Board or member countries concerning the legal status of any country, territory, city, or area or of the authorities thereof or concerning the delimitation of its boundaries or its national affiliation.

Because of the informality and to present the results of research with the least possible delay, the typescript has not been prepared in accordance with the procedures appropriate to formal printed texts, and the World Bank accepts no responsibility for errors.

The material in this publication is copyrighted. Requests for permission to reproduce portions of it should be sent to Director, Publications Department, at the address shown in the copyright notice above. The World Bank encourages dissemination of its work and will normally give permission promptly and, when the reproduction is for noncommercial purposes, without asking a fee. Permission to photocopy portions for classroom use is not required, though notification of such use having been made will be appreciated.

The complete backlist of publications from the World Bank is shown in the annual *Index of Publications*, which contains an alphabetical title list and indexes of subjects, authors, and countries and regions; it is of value principally to libraries and institutional purchasers. The latest edition is available free of charge from the Publications Sales Unit, Department F, The World Bank, 1818 H Street, N.W., Washington, D.C. 20433, U.S.A., or from Publications, The World Bank, 66, avenue d'Iéna, 75116 Paris, France.

Guy Le Moigne is chief of the Agricultural Production and Services Division of the World Bank's Agriculture and Rural Development Department, Shawki Barghouti is the department's adviser on irrigated crops, and Herve Plusquellec is the department's adviser on irrigation engineering.

Library of Congress Cataloging-in-Publication Data

```
Technological and institutional innovation in irrigation.

   (World Bank technical paper, 0253-7494 ; no. 94)
   "Proceedings of a workshop held at the World Bank,
April 5-7, 1988."
   1. Irrigation--Technological innovations--Congresses.
2. Appropriate technology--Congresses.  I. Le Moigne,
Guy.  II. Barghouti, Shawki M.  III. Plusquellec,
Herve L., 1935-     . IV. World Bank.  V. Series.
TC809.T43  1989          627'.52          89-5732
ISBN 0-8213-1185-9
```

ABSTRACT

This volume is a compendium of papers presented at the "Workshop on Technological and Institutional Innovation in Irrigation" held in Washington, D.C. in April 1988, and sponsored by the World Bank. The main objective of the workshop was to examine the available technologies in irrigation and to ask whether they have been adequately and effectively exploited throughout the world -- particularly in developing countries. In the first section a review was made of changing roles in irrigation and of technology and research issues in irrigation. The conclusion of this section is that irrigation research has not kept pace with agricultural scientists who have dramatically changed the potential yields that can be obtained from good crop husbandry. The second section focused on the choice of appropriate technology. Non surface methods dominate the discussion primarily because the main advances in irrigation over the last few decades have been in sprinkler and drip techniques.

The third section reviewed research priorities. More research is urgently needed across a very broad front, above all the interface between agriculture and hydraulics in surface irrigation, and economic, social and institutional aspects of modern irrigation systems.

The fourth section focused on transfer of technology. Significant effort should be made through seminars, training programs, and the use of consultants and international and local experts to promote the transfer of technology. Also, effort should be made to coordinate and promote international research and transfer of technology by organizations such as the International Commission on Irrigation and Drainage.

CONTENTS

Part I

OVERVIEW

Part II

CHOICE OF APPROPRIATE TECHNOLOGY

Part III

RESEARCH PRIORITIES

Part IV

TRANSFER OF TECHNOLOGY

PREFACE

Are surface irrigation systems behind the times? Has irrigation technology stagnated while agricultural production was radically transformed by the advent of the green revolution? These questions were among several important issues addressed during a unique workshop organized by the World Bank on "Technological and Institutional Innovations in Irrigation."

The importance of irrigation cannot be overemphasized in our efforts to develop agriculture. In Asia the irrigated areas, although covering less than half the cultivated land, provide almost two-thirds of the food production. About one-third of the population of the world depend directly on irrigation for food. To meet the future growth in demand for food and fiber there will be a need for substantial new investment in irrigation. In most of the vast areas of traditional irrigation the systems retain the design concepts of the early part of the century -- both in the new and the old projects. There is an urgent need to review the old designs and research new concepts that will respond better to the demands of modern agriculture.

In a few countries, the traditional canals have been replaced by more efficient piped systems that save water and reduce the problems of drainage and salinity. During the workshop, the participants reviewed the scope for improving the traditional irrigation systems and analyzed the reasons behind the success of modern pressurized irrigation. The participants concluded that: (i) while in the past, simple extension of water supplies to embrace more farmers with speedy and cheap irrigation water was a laudable and effective objective, future demand on water would require more efficient and modern irrigation methods; (ii) shortages of land and water resources constrain opportunities for expansion of surface irrigation in many developing countries; and (iii) farmers are moving to take advantage of the agricultural advances that offer higher production and incomes through intensification and diversification. The proceedings of this workshop provide first hand assessment of how this process is taking place in several countries, especially the USA, Israel and Jordan.

The discussions during the workshop of the remarkable experience of these countries in modernizing irrigation systems and the scope and limitations for expanding this experience to other countries are well presented in this volume. The first part of the proceedings provides us with an articulate summary of the issues on which the workshop focused. In the second part, we are presented with detailed discussions of the country experiences from the Middle East, especially Israel, the United States and France. In the third part, research issues and priorities are presented with a practical reference to what is actually taking place in selected countries. The last part focuses on issues related to technology transfer, with a special attention to the role of the private sector and public institutions dealing with research and extension. The proceedings published here should provide an excellent reference on the subject of how to develop and improve irrigation systems.

Valuable contribution to this debate was made by expert representatives from India, China, Brazil, Morocco, Pakistan, Israel, Jordan, United States, United Kingdom, France, the US Agency for International Development (USAID), the Inter-American Development Bank (IDB), and the World Bank.

Mr. Rajagopalan supported the initial concepts and objectives of the workshop and requested Mr. Le Moigne to organize it. The joint effort between AGRPS and EMENA had made this workshop successful. Several people have contributed to this effort including H. Plusquellec, S. Barghouti, P. Bielaski and D. Duggin.

Michel Petit, Director
Agriculture and Rural Development Department

PART I

OVERVIEW

Chapter 1

THE WORLD BANK'S CHANGING ROLE IN IRRIGATION

V. Rajagopalan

The World Bank has played a significant role in financing irrigation and drainage projects for almost forty years. Since 1950, it approved loans to 317 irrigation and drainage projects in approximately fifty countries. The Bank provided almost $14 billion for these projects, whose total costs are estimated to be $32 to $33 billion. In the 1970s, there was considerable interest in the subsector, and 80 percent of these projects were approved since 1973. Recently, the Bank has decreased its lending for such projects. The current portfolio is comprised of 173 irrigation and drainage projects, most of which are in Asia (63 percent); 24 percent are in the EMENA Region, 9 percent are in Latin America, and fewer than 2 percent are in Africa.

There are approximately 220 million hectares of irrigated land in the world. Over 60 percent of this area is in five countries--China, India, the United States, the Soviet Union, and Pakistan. More than two-thirds of the world's total irrigated area is in developing countries, and of this area, 60 percent is in India. The area brought under irrigation has increased rapidly since the early 1960s; approximately 80 million hectares were added. These projects have resulted in a remarkable increase in the production of rice, wheat, and other crops. Several countries that had cereal deficits have recently become self-sufficient in cereal grains. Several countries are projecting a doubling of their irrigated areas by the end of the century.

Despite these impressive increases, there is a general concern that irrigation systems are performing well below their potential, and the following problems have been identified:

> Overall efficiency of water use may be 30 percent or less, whereas well-managed systems show efficiency of 50 percent or more.

> Inequalities in the pattern of water distribution to farmers are common, resulting in excess water in some places and deficits in others. When the water supply is insufficient or uncertain, farmers at the head of the system take advantage of their location and use an unfair share of the water.

> Waterlogging and salinity problems have developed in arid regions. Approximately 40 million hectares of irrigated lands are estimated to be affected by salinity.

Because of poor water management, benefits from many irrigation projects are far short of expectations. As a result, irrigated areas are smaller and crop yields are lower than estimated at the planning stage.

When the water supply is unreliable, farmers in irrigated areas are not motivated to organize themselves and participate in operating and maintaining the tertiary distribution network. When the service is poor, farmers are less willing to pay the service charges.

Operational problems are aggravated by deferred maintenance that results from inadequate funds. Lack of maintenance leads to deterioration of the assets.

None of these problems is new, and the need for improving the performance of irrigation and drainage systems is well recognized. Many countries are already facing water shortages, and allocation of available water resources among competing requirements is becoming a difficult task. Many irrigated areas in the world suffer from waterlogging and salinity, which is often induced by excessive water use and poor drainage facilities. Many projects do not have the flexibility and reliability of water distribution required for crop diversification and the cultivation of high-yield varieties. Achieving food security is an important objective of many countries, but to attain it, some countries will have to improve and modernize irrigated agriculture. Many of these systems were built in the late nineteenth century and were suited to conditions at that time.

The World Bank's approach in many areas has changed since 1950; its role in irrigation and drainage has changed especially since the mid-1970s. The Bank financed the construction of more than 350 single- or multi-purpose large dams for irrigation, power, and water supply. Fewer dams are now being built. In addition, the Bank and other aid agencies in the early 1960s financed the construction of the main distribution systems associated with the dams and assumed that farmers would take care of the distribution systems in their own areas. That effort was not forthcoming as anticipated, however. Many types of on-farm development, including land leveling, farm irrigation, ditches and drains, farm roads, and even land consolidation in areas where local authorities are convinced of the advantages of a geometric layout, are now components of Bank-financed irrigation projects.

Since the mid-1970s, the Bank has progressively shifted from developing new land for irrigation to rehabilitating or modernizing existing irrigation projects. More recently, to improve the performance of existing irrigation schemes, the Bank has financed a new generation of irrigation and drainage projects in which management and policy issues are the focus. One of these projects is the recently approved National Water Management Project in India, which is initially confined to three states in South India but could be expanded as it becomes successful. The project was an outgrowth of pilot projects in India which demonstrated that the productivity of a large irrigation system could be improved by changes in water scheduling. The water is delivered according to clearly defined rules that are designed to meet the agronomic requirements of the predominate (and evolving) cropping pattern but do not attempt to meet the individual needs of innumerable farmers. With the management task made very clear, the system is easier to control, and farmers can plant their crops and arrange their farming activities to fit the pattern of water supply. The water supply is more reliable and predictable, although it is not absolutely guaranteed.

The Irrigation Operations Support Project in the Philippines is a second example of this new generation of projects. In this case, the objective is twofold: to improve the operating performance of the national irrigation schemes through minor rehabilitation works and increases in the annual funding for operation and maintenance (O & M) services; and to strengthen substantially the institutional capability of the National Irrigation Authority and private irrigation associations to improve and maintain the efficiency of existing infrastructure. The goal of the project is to stop the decline in the irrigated areas; no increase in yield is expected.

In both of these projects, the cost is small--about $140 per hectare in India and $75 per hectare in the Philippines. The physical components of these projects are also relatively minor; they include maintenance works, desilting of canals, local repairs, and construction of a few new structures designed to simplify operation of the schemes. The main innovations are in the area of management, namely, the transfer of responsibilities for O & M to farmers and the improvement in water management and collection of service fees.

To help solve the problem of poor performance of irrigation projects, the Bank has focused recently on several areas of management: improved water resources management, better coordination between irrigation and agriculture agencies, better training, larger budget allocations for operation and maintenance, higher water charges, and improved farmer participation. These measures have not always appreciably improved the situation. In some rehabilitation projects, repairs have been required only a few years after completion of the initial project, which suggests that changes need to be made in the fundamental design of the system. Commenting on inherent design and management limitations, two experts from the International Irrigation Management Institute cited experience in South Asia in a recent publication:

> What will be the impact of better system maintenance upon agricultural productivity? Probably, not much. Low productivity is unlikely to be corrected by doubling O & M expenditures. O & M preserves the capacity of the irrigation system as it was designed and is currently used. Increased investment in O & M doesn't address inherent design and management limitations, which along with a lack of agricultural inputs, more directly constrains agricultural performance. Thus, while better O & M is needed and increased financial support is justified, neither is a substitute for improvements in design and management. . . .

> The continuing implementation of water course rehabilitation programs, even including recent efforts to organize water-users associations will be problematical from a long-term perspective in the absence of more

fundamental changes in system design and the organization of irrigation-related tasks.

The remarkable technical advances in irrigation have solved many problems, but many issues must yet be resolved. Advanced and high-technology systems, particularly in the EMENA Region, were financed by the Bank over the years. A variety of approaches to irrigation are evident--from spate irrigation in Yemen and Afghanistan to drip irrigation in Cyprus. The Bank financed sprinkler irrigation systems in Yugoslavia, Romania, Morocco, and Tunisia. Some developing countries have adopted the most advanced technology, such as remote centralized control. In the Persian Gulf, countries in which water is extremely scarce have begun to use drip irrigation systems and greenhouses.

In addressing the question of how to improve the performance of irrigation and agriculture, several factors should be under consideration. The need to modernize existing systems is clearly evident, but the appropriate choice of technology in developing countries is an important issue. Should high-technology systems be considered in these countries? Should we continue with the current practices of design or move to more innovative techniques and institutional solutions? How can the advanced techniques used in developed countries be adapted to suit the socioeconomic conditions and the human resource availability of developing countries? What is the appropriate combination of technical and institutional innovations?

If the proportion of World Bank lending in the agricultural sector continues to remain at approximately 25 percent in the next few years, it is essential that answers to these questions be found. These issues need to be addressed not only in the context of on-farm facilities and farming activities but also in the context of management of the entire irrigation system. The papers presented at the workshop, which focus on three themes--technical innovation, research priorities, and conditions for transfer of technology--make a valuable contribution to shaping the Bank's strategy for assisting developing countries in improving irrigated agriculture.

Chapter 2

OVERVIEW OF TECHNOLOGY AND RESEARCH ISSUES IN IRRIGATION

Guy Le Moigne

One of the main objectives of the workshop was to examine the available technologies in irrigation and to ask whether they have been adequately and effectively exploited throughout the world--particularly in developing countries. According to one school of thought, the irrigators have not kept pace with the plant breeders and others who have dramatically changed the potential yields that can be obtained from good crop husbandry. Is it in fact true that we irrigators have lagged behind our agricultural colleagues over the last few decades? Several answers to that question can be found in the papers presented at the workshop, which were devoted to three main issues: choice of irrigation technology, research priorities, and transfer of technology.

Many of the papers focus on drip and overhead methods of irrigation, an emphasis that comes at the expense of discussing technological advances in surface irrigation, which constitutes about 85 percent of the irrigation systems in the world. Non-surface methods dominate the discussion primarily because the main advances in irrigation over the last few decades have been in sprinkler and drip techniques. Furthermore, research and development in irrigation have come mainly from the private sector--from the manufacturers of equipment. The private sector has had only a limited interest in surface irrigation because there is very little marketable hardware in it.

The seminar participants reported some of the important advances that have taken place in drip and overhead irrigation. Of particular interest is the impressive way in which the latest technologies have been adopted in countries of the Mediterranean and in Brazil, China, and elsewhere. Drip and sprinkler methods are particularly well suited to the production of high-value horticultural crops in regions where labor is expensive and water resources are scarce and on uneven land and soils having high infiltration rates. Many of these conditions are found in the Mediterranean countries; it is not surprising that drip and sprinkler methods have expanded rapidly in this region. There are many other regions in the world particularly well suited to drip and sprinkler methods: Brazil, Mauritius, parts of China, and the United States. In the immediate future, drip and sprinkler techniques are likely to be applied in selected areas in most parts of the world, including Asia, wherever the specific conditions favor such methods.

In surface irrigation there have been some significant advances in the use of pipelines for water conveyance, but these improvements have occurred mostly in smaller systems. For the vast systems of Asia, the available technologies remain much the way they were more than fifty years ago. These larger systems have encountered some major problems. Almost all of them are based either on a form of continuous flow or on a turn system that depends on a fixed discharge. With the notable exceptions of the Sudan,

Egypt, and Morocco, there are few systems that provide water supplies against "indent" or demands. While inflexibility of supply in the vast Asian systems had previously caused few difficulties, severe problems now occur as cropping intensities increase and farmers need more reliable water deliveries to meet the requirements of modern crop varieties and other farm inputs. Furthermore, the intensification of cropping without better water control aggravates the drainage problem and thus increases the incidence of waterlogging and salinity. It is doubtful whether crop production can be further intensified in existing irrigation schemes without the introduction of much better methods of water control. Crop diversification in the traditional rice-growing areas of Asia requires some radical departures from the old continuous-flow system of water delivery.

It is important to consider the reasons for the apparent technological stagnation in surface irrigation. There is clear evidence that little progress in surface irrigation technology has been made over the last few decades. Perhaps standards have declined as management organizations have become overwhelmed by pressures for more intensive cropping and hence more water. But on the brighter side, a dramatic rate of expansion of total irrigated area has taken place. In the two decades from 1950 to 1970, gross irrigated area doubled, and over the second half of this century it will probably be trebled. Many countries have consciously pursued a strategy to expand irrigation as rapidly as possible using low-cost (and low-efficiency) approaches. The primary aim has been to provide more irrigable land as quickly as possible, although that meant facing the problems of modernization later. This strategy might be dangerous because modernization of existing schemes is always difficult. Several speakers pointed out that as systems are developed, people tend to become locked into them and show unwillingness to adopt change unless they are faced with clear and perhaps dramatic failures. In surface irrigation, there is an urgent need to take stock of the situation and start developing new systems that are well suited to the requirements of modern agriculture and will permit the achievement of sustained yield levels much nearer to the potential yield of the new crop varieties.

It is not an exaggeration to say that we do not know how to deal with many of these problems. More research is urgently needed across a very broad front, an effort that includes all the associated factors--economic, social, institutional, and technical. Above all, the interface between agriculture and hydraulics in surface irrigation needs further research--as has been done in non-surface irrigation methods. The World Bank supports research in many areas both directly and through its project-lending programs. More recently, it has given substantial financial support to the International Irrigation Management Institute, a research organization focused on the management and institutional factors of production.

Whether the Bank should do more in supporting irrigation research has been a question put by President Lahlou to Mr. David Hopper, the Senior Vice President responsible for policy and research planning in the Bank. The International Commission on Irrigation and Drainage (ICID) asked the Bank to take the initiative in creating an international support group to increase and enhance irrigation research in developing countries. In response to this request, several steps have been taken. The Bank has proposed first to review the status of irrigation research in the world, making an inventory of ongoing research topics and assessing the more urgent needs. On completion of that review, it is hoped that several countries will be interested in creating a support group. It is too early to judge the form that group might take, but more information should be available by the time of the next meeting of the Executive Council of ICID in Yugoslavia in September 1988.

The Bank also could do more to assist in the transfer of technology. While it is already actively engaged in this effort through seminars, training programs, the use of consultants and the Bank's own experts, and through the support of international organizations such as the ICID, the Bank could support the transfer of technology by coordinating and promoting a network of international research. This matter and the appropriate institutional arrangements will be under further consideration.

An underlying question, one that is dealt with in Chapter 2, is whether the Bank should play a larger role in raising technological standards in irrigation. It is possible that the Bank has followed, rather than participated in, the development of new technologies. This question deserves careful consideration, and I hope that in the future we shall be able to come up with useful initiatives to help developing countries form new concepts and remodel old systems along modern lines and thus improve the quality and efficiency of irrigated crop production.

PART II

CHOICE OF APPROPRIATE TECHNOLOGY

Introduction

Choice of appropriate technology in irrigation is the topic of the papers presented in this section, which describe water application technologies in current use in three developed countries, the United States, France, and Israel, and in selected countries in the Middle East and North Africa. They point out some examples of failures of inappropriate technology.

Technologies vary greatly within the same country or region. As the cost of water increases, the expansion of irrigation becomes less feasible, particularly where labor and other input costs are also rising. On the other hand, as markets for high-value crops expand, more efficient technologies are adopted. Many of these factors interact; an evaluation of them in light of established practices elsewhere under similar conditions is likely to indicate the pattern of technological development in a comparable setting.

At every level of economic development, surface irrigation predominates in all countries that have large irrigated areas. The more sophisticated surface methods available today have yielded results in line with farmers' general criteria for maximum net income. Under normal market conditions, two factors, land irrigability and water scarcity (reflected in its cost), have the greatest impact on the choice of technology. Some specialty crops, however, dictate the choice of technology because their cultural requirements may be significantly different from those of most other crops under the same conditions. Disease and environmental setting may also inhibit free choice.

In virtually all cases, farmers adopt the technology that produces the greatest net income within acceptable levels of risk and complexity. Evidence from the World Bank's post-project evaluations and experience in various countries has shown the disastrous results of forcing the development of unjustifiably high technologies. What is essential to rapid adoption of a new technology is farmer exposure to it through neighbors, demonstration farms, and sound support from private-sector suppliers. Farmers with low tolerance of risk will remain cautious about investing their money and effort into new methods.

A range of irrigation technology exists today with well proven attributes. A great amount of practical information is readily available from many sources on operational characteristics, performance, and costs of specific technologies. Project plans and on-farm results demonstrate, however, that much of this information is still not widely utilized.

While World Bank loans for irrigation and drainage, which currently amount to $14 billion, continue to constitute a substantial portion of the total loan portfolio, the Bank's emphasis has been shifting from loans to projects in which dams and large works predominate to investment in delivery systems, on-farm works, and operation and maintenance. Water application technologies are of increasing concern (see Chapter 2).

In Mr. Van Tuijl's presentation, he described some of the technologies found in the Bank's region of operation in Europe, the Middle East, and North Africa (EMENA). Two examples represent the extremes in technology choice: brush diversion dams, annually rebuilt by the water users on old remote systems in Afghanistan; and piped service to drip systems and sprinklers in Cyprus. Problems of waterlogging and salinity are increasingly encountered in countries such as Turkey, Pakistan, and Egypt. No apparent major shifts in technology are likely; rehabilitation of existing systems is becoming the major need in the region.

Mr. Melamed outlined the Israeli experience with the evolving irrigation technology. The adoption of irrigation methods is dictated by the particular set of conditions that prevail in Israel: limited, costly water supply, high labor costs, high-value crops, and difficult lands. Implementation has been facilitated by intensive government involvement and a strong private equipment service sector. Sprinkler and drip systems predominate, particularly on permanent and high-value crops; little area is served by surface methods. Automation is increasingly a feature of low-volume, high-technology systems.

Mr. Burt presented extensive information, much of it derived from ongoing joint university-government field studies, on the numerous variations found with surface, sprinkler, and drip systems in use in California. Surface methods still predominate in California (79 percent), but with such improvements as gated pipe and pumped tailwater systems, the efficiencies obtained approach those of more sophisticated methods. Drip irrigation is limited to marginal land with high-value crops, primarily vine and tree crops, and is no longer expanding because of costs, operational difficulties, and expanded choice among micro sprinklers. A range of sprinklers, tailored to the situation, is used extensively on both field and tree crops. Automation is minimal, since the operational advantages and reliability of equipment have not proved to be sufficient. Only 4 percent of farmers use Penman-based water scheduling. But the practice of administering fertilizer through irrigation water is increasing. Net income of the entire cropping operation is what determines the selection of technology. A flexible, reliable supply of water to farmers is the key to allowing them fully to utilize available technologies to meet particular conditions.

Mr. Manuellan described the experience with application technologies in France. The change from surface irrigation to sprinklers began in the 1960s and 1970s, as rising labor costs generated great interest. Increasing energy costs contributed to the faster growth of drip irrigation over that of sprinklers. Automation gained greater prominence in the 1980s as labor costs rose even further. Over the last few years, attention has shifted, and once again, interest in surface irrigation has grown. The focus is now on adapting new methods and equipment to the old surface systems. The supposedly "best" technology is not always utilized by farmers, sometimes for lack of proper introduction. In all cases in which equipment is required, the key is an aggressive, strong, supporting private sector.

Discussion of the choice of technology papers by the participants confirmed much of what the speakers had found. A reliable, timely water supply is an essential starting point for using any improved technology. Surface application methods will continue to dominate, warranting a proportionate amount of attention during planning and implementation. Sprinklers have greatest acceptance on difficult lands; 55 percent of lands irrigated in the United States outside of California are now sprinkled. If farmers' objectives for net return are met, the most significant factors in adoption of technology are farmer training and private-sector support.

Chapter 3

IRRIGATION DEVELOPMENTS AND ISSUES IN EMENA COUNTRIES

W. A. Van Tuijl

Irrigation practices in the EMENA region vary widely. In some countries, one can find both traditional and modern methods--irrigation systems that have been crudely built and operated by farmers and modern canal or pipeline systems with on-farm sprinkler and drip systems. This paper begins by describing the most traditional methods and progresses to the most advanced systems.

Farmer Built and Operated Systems

In several of the EMENA countries, particularly Afghanistan, the Yemen Arab Republic, and the People's Democratic Republic of Yemen, irrigation systems built and operated by farmers are in evidence, and such systems can still also be found in Morocco, Algeria, Tunisia, Egypt, and Pakistan. Although some schemes take water from perennial rivers, in most cases the water supply is seasonal. Spate irrigation--a system in which the main sources of irrigation water are spate flows of short duration--is of particular importance in the arid regions of the Arabian peninsula where catchments are rocky and steep. Based on centuries-old experiences of farmers, these traditional methods are elementary but effective. The deflectors or earthen banks built by farmers across the channel of the *wadi* divert the spate and base flows to the fields. Large floods usually wash away these diversions, thus reducing or preventing irrigation of the fields. Although they are cheap to build, they have high seasonal maintenance costs. Depending on their location in the system, farmers might irrigate several times a year or only once in several years. Greater stability in agricultural production would require better control of spate flow and improved irrigation facilities.

Improved Spate Irrigation

Since the mid-1960s, several *wadi* development projects have been designed and implemented, but the results of the often very capital-intensive improvement works have been variable, which reflect the complex nature of spate irrigation structures. During the feasibility stage, often the most important issue to be decided is whether to build a storage dam or one or more permanent diversions. The structures are generally expensive because they have to withstand high floods and must be capable of handling a large quantity of sediment. More innovative approaches, such as the use of gabions, mattresses, and fuse plugs, have been followed recently, which reduce the cost of diversion structures. In December 1987, FAO organized a regional conference on spate irrigation in Aden, whose participants concluded that several issues required particular attention: the need for more detailed studies of traditional water rights and for priorities and wishes from farmers in improving their irrigation systems; the economic aspects of agricultural production and the potential for its improvement; project design based on systematic collection of basic climatic, hydrological, and hydrogeological data; the choice of irrigation, sediment-control, and flood-protection works through careful analysis of costs and benefits; improved research in spate irrigation crops; the integration of surface and ground water in optimizing crop production; and the need for farmers to share responsibility for operation and maintenance.

Pakistan: A Traditional System Designed for Drought Protection

Traditional large-scale systems in Pakistan irrigate about 16.2 million hectares. Irrigation is a necessity for most agricultural production, since precipitation is less than 150 millimeters a year and pan evaporation varies from 1,250 to 2,800 millimeters. The quality of the surface water is good for irrigation. Of the extensive ground water underlying the Indus Plain, nearly 50 percent is of satisfactory quality for irrigation. Fresh ground water pumpage supplies about 42 billion cubic meters a year for irrigation. About 38 percent of the lands are either saline or sodic, or a combination of the two, and 22 percent have a water table within 2 meters of the surface, with an additional 20 percent having a water table within 3 meters. The yields of major crops have remained well below those of many other developing countries. In 1983, wheat and paddy yields in Pakistan were only about 1.6 tons and 2.5 tons per hectare, respectively. Cropping intensity averaged 96 percent in 1985.

Pakistan's systems were designed to provide initially only for irrigation facilities; drainage facilities were to be provided when needed. The design criteria were tailored to several factors: the availability of water supplies; the objective of bringing crops to maturity on the largest possible area with minimal consumption of water; and low operating costs and limited technical staff.

Water distribution is characterized by main and branch canals that supply water to a large number of distributries, which generally run continuously; the canals run by rotation when demand for water is less than about 50 percent of peak demand or when availability is restricted. The distributries supply water to water courses through constant flow modules. The water is allocated to the farmers on a water course by a time roster (Warabandi system). The basic principal of the Warabandi system is that each farm, according to its size, receives water on the same day each week for the same length of time.

Overall, system efficiency in Pakistan is very low, about 45 percent. Conveyance efficiency is 75 percent, and efficiency within the irrigation unit (*chak*) is 60 percent. Net crop use on the farm, including rainfall and deliveries from ground water, average only about 645 millimeters, which is far below potential evapotranspiration. Even with the addition of the Kalabagh Dam in the future, this figure, on average, would not increase to more than about 700 millimeters.

In addition to heavy losses, the systems also suffer from inequitable water supply, which is more pronounced between *chaks* than within the individual *chaks*. This variability is largely because of sedimentation in the upper reaches of the canals. In theory, the time periods allocated to individual farmers in the Warabandi roster should be adjusted for the travel time of the water in the water courses and for seepage, but in practice, this is not done. There is allowance for travel time but not for seepage. Within the *chak*, tail-end farmers receive less water than do farmers near the outlet.

The irrigation system is generally in great need of rehabilitation. Freeboard of canals is often inadequate, embankments and structures have deteriorated, and embankments and canal cross sections have been ravaged by unauthorized crossing of people, animals, and vehicles. In addition, the number of escapes to evacuate excessive flow from monsoon rains is inadequate.

Because Pakistan's irrigation systems will continue to operate under scarcity conditions, there is general consensus that the present method of system operation and the Warabandi system should be maintained. A "revised action program for irrigated agriculture" was prepared in May 1979, which recommended numerous improvements in policy and institutions, namely, privatization of public tubewells in freshwater areas. Investment in drainage, rehabilitation, and water management should be of high priority. These conclusions and recommendations are still valid. A second rehabilitation project is under preparation for Bank financing. The strategy for rehabilitating the irrigation systems includes the following: rehabilitation of the main system; rehabilitation of the water courses, including about 15 percent lining in fresh ground water areas and 30 percent lining in saline ground water areas; and improved operation and maintenance. Investment in public works is estimated at about 800 rupees per hectare (equivalent to US $45 per hectare).

Turkey: A Conventional Irrigation System

In Turkey 17 percent of all arable land is irrigated. About 40 percent of all crop output and about 25 percent of agricultural exports are grown under irrigated conditions. But in many respects, the performance of irrigated agriculture has been considerably below its capacity. Newly developed irrigation projects have often failed to produce expected increases in production; the increase in irrigated-crop area has lagged significantly behind the expansion of irrigation infrastructure. This underutilization of infrastructure, which was built at comparatively high cost, has resulted in a large loss to the economy.

Of the 28 million hectares of arable land, about 4.7 million hectares have been brought under irrigation. Of this total, major government projects irrigated 1.3 million hectares, small irrigation schemes constructed by the government irrigated 1 million hectares, and privately developed irrigation totaled 2.4 million hectares. Cropping intensity is low, only 83 percent, largely because of the considerable lag between completion of basic infrastructure and completion of drainage and on-farm works. Equipped areas have failed to reach their full cropping potential, and production has actually fallen over the years because of waterlogging and salinization. Furthermore, about half the irrigated areas are in regions with cold winters, and cultivation is limited to one crop. By removing the physical constraints, however, it should be possible to attain a cropping intensity of about 120 percent.

Many schemes remain uncompleted because two public agencies have been responsible for the construction of irrigation facilities. The General Directorate of State Hydraulic Works (DSI) of the Ministry of Public Works and Settlement is responsible for the construction of the basic irrigation infrastructure, including drainage and flood protection provided by canals commanding 2,000 to 3,000 hectares. The General Directorate of Rural Services (GDRS) of the Ministry of Agriculture, Forestry, and Rural Affairs is responsible for on-farm and subsurface drainage, land leveling, construction of water courses, rural roads, domestic water supply, etc. Unfortunately, coordination between the DSI and the GDRS has not been satisfactory for planning and funding of new schemes, and there has been a lack of cooperation in technical matters, which has resulted in occasional technical deficiencies and a considerable lag between completion of basic irrigation infrastructure and actual irrigated area--about 266,000 hectares in 1984.

Most of Turkey's irrigation schemes rely on storage dams for their water supply. The canal distribution systems in new schemes are generally lined. In command areas in undulating terrain, extensive use has been made of elevated concrete flumes (canalettes). Water-control structures are traditional. Large structures may have sector gates, whereas structures on small canals and turnouts have slide gates. In some schemes, constant head orifices (CHOs) have been provided, but generally they are not used. Water is delivered to irrigation units of 25 to 40 hectares, and within these units farmers take care of their own water distribution. Water is delivered to irrigation units 24 hours a day, which leads to waste during the night, but this is not perceived as a problem since water availability in Turkey, especially in coastal areas, is more than adequate for meeting agricultural needs. This waste of water, along with the deterioration of surface drains because of lack of maintenance, however, has led to waterlogging and decreased production (reduced yields and cropping area). The area affected by poor drainage has been estimated at about 450,000 hectares. The government acknowledges that something needs to be done to rectify the situation, and its investment programs gives high priority to completion of irrigation infrastructure, provision of drainage infrastructure, and improvement of operation and maintenance. In addition, improved extension will help achieve the government's objectives of crop intensification and higher yields.

Egypt: A Traditional Irrigation System with Low-Level Canals

Agricultural production in Egypt almost entirely depends on irrigation with water from the Nile, as the amount of rainfall is negligible. Approximately 6 million *feddans* (2.4 million hectares) are irrigated; of that total, 5.5 million *feddans* are in the traditional Nile valley and delta, and the remainder are lands reclaimed from the desert (part of the 0.9 million *feddans* nominally reclaimed). Egypt's land reclamation effort has not been very successful, and such schemes have been estimated to contribute not more than 3 percent of the total value of agricultural production.

Egypt has usually enjoyed an abundant supply of water since the completion of the Aswan Dam. Under its agreement with the Sudan, Egypt is entitled to 55.5 billion cubic meters a year, although actual usage has reached 57.5 billion cubic meters. But recently, as a result of the continuing Sahelian drought, the water level in the Aswan reservoir was predicted to reach dead storage by April 1988, and Egypt may have to reduce its annual releases from the reservoir. The main issue is the duration of the Sahelian drought; there seems to be a growing consensus among leading meteorologists that it will continue. The continuing drought would stop the implementation of additional land reclamation schemes. Even with the original flows restored, however, the main increases in Egypt's agricultural production would have to come from the so-called "old lands."

The Egyptian irrigation system suffers from several deficiencies. Structures are broken and deteriorated. Water distribution is inequitable, which is a result of inadequate discharge control. Discharge takes place only at the head of the nine governorates, and below this level, there is only control of water level and not discharge. Occasionally there are tail-end problems because of constraints in the system--notably in the case of long and flat *mesqa* (water courses). Operation of the canals is on a rotation basis, but actual deliveries often differ from those planned. The system has worked remarkably well, however, judging from the average cropping intensity of over 190 percent and crop yields per hectare of 3.3 tons for wheat, 4 tons for maize, 3.8 tons for sorghum, and 5.5 tons for rice. System efficiency is about 65 percent on average, with a peak of 83 percent in July and a low of 40 percent in winter. The system has worked well for several reasons. Although the system is operated from "top to bottom," farmers receive water in the *mesqa* typically 0.5 meters below the elevation of their fields and lift it to their fields using animal-driven *sakia* (waterwheels) or diesel-powered pumps, which reduces operational waste. The low-level canal system provides night storage and compensates for the mismatch between supply and demand. Considerable amounts of drainage water are reused, through public pumping stations and direct pumping by farmers.

A challenge for Egypt is to undertake a low-cost rehabilitation of the existing system that would allow it to operate under future conditions of scarcity, to make the water supply more equitable, and to improve overall operating efficiency by further reducing losses to the sea. Studies have recently been completed under a United Nations Development Planning (UNDP) project, for which the World Bank acts as executing agency. Under this project, command areas totaling 150,000 *feddans* were studied in detail, and general criteria were developed for the rehabilitation and improvement of the existing systems. The study made several recommendations. The present advantages of the existing system, such as low-level canals and low-lift pumping, should be maintained. A changeover from rotation between canals to a continuous flow system should be implemented; the water supply to farms would thus be more dependable, more effective use would be made of canal storage, and the constraints on capacity could be overcome. Appropriate "intervention points" should be identified between the upstream system with scheduled operation and the downstream system with demand operation, and discharge measurements should be provided at these points. Reuse of drainage water should be further developed. Ground water should be used to overcome tail-end shortages in areas wherever feasible. Among other things, *mesqa* improvements should to made to eliminate losses from leaking collector-drain crossings and outlet pipes to drains. The cost for rehabilitation has been estimated, in 1987 prices, at $920 per hectare.

The selection of appropriate intervention points depends upon many factors, including the characteristics of the existing layout, organization requirements, and farmers' willingness and ability to cooperate. Presently, command areas that are supplied independently vary in size from 5,000 to 15,000 *feddans*, but they will be reduced to irrigation blocks of about 2,000 to 6,000 *feddans*. The water-control structures envisaged for the rehabilitation of the 150,000 *feddans* include the following: 3 radial gates, 58 distributors, and 15 automatic downstream water-level control structures. For the most part, distributors have been proposed; they can provide both cross regulation and discharge control. These structures will be useful especially in intermediate reaches of canals that receive a constant discharge at the head but also have to pass a constant discharge at the downstream end to another irrigation block. The distributors would be designed to allow for the planned night storage range but with provision for overflow should levels exceed this range.

Morocco: Surface Irrigation with Automatic Hydraulic Control

Advanced irrigation technology has been the focus of various World Bank seminars and an irrigation study tour that took placed in 1985. While the focus here is on innovation in Morocco, it should be mentioned that its conventional systems have many of the disadvantages of those in Turkey. Overall irrigation efficiency is low--sometimes as low as 20 to 30 percent. The intensity of irrigation is low. The water supply is unreliable and inequitable, and as a result, farmer participation in O & M has been ineffective.

Hydraulic automation, including simple devices such as duckbill weirs, has been in use in Morocco on a large scale since the 1950s. While automatic upstream control gates are used, there is still a need to establish a program of water supply to branch canals and to schedule releases at the headworks in advance. Operational losses generally cannot be reduced below 10 percent. With automatic downstream control gates in a main canal, for example, changes in branch-canal requirements are automatically transmitted to the head of the canal, and the discharge in the main canal is automatically adjusted to match the overall demand. Composite gates have also been developed, which combine the advantages of upstream and downstream control. Most schemes have a combination of downstream control gates in the larger canals and upstream control gates in secondary canals. The discharges at the heads of main canals are controlled by the project authority. Automatic control gates are often used in conjunction with calibrated distributors or "modules" for discharge measurement at the head of canals or irrigation units. These gates can be preset for specific discharges and are relatively insensitive to variations of plus or minus 5 percent in upstream water levels. The Doukkala project in Morocco incorporates these devices. The conveyance and distribution efficiency of the gravity commanded area (about 27,500 hectares) has reached about 76 percent. The cost of the distribution system in 1986 prices was US $3,040 per hectare.

Morocco has initiated two institutional innovations that are intended to improve efficiency. The first objective was to improve the efficiency and financial autonomy of the nine Regional Agricultural Development Authorities (ORMVA) through increased water charges.[1] The second objective was to improve operation and maintenance of existing large-scale irrigation systems covering about 400,000 hectares. The institutional changes are comprised of several elements. Each ORMVA would prepare detailed three-year development plans that would define the agency's medium-term objectives and operating targets. These plans would be the basis for program contracts between the ORMVAs and the government. The development programs would identify the objectives of the ORMVA in terms of the following: investment; field activities, such as the amount of water distributed, the number of farmers contacted, types of system activities to promote production, and so on; targets for new management-related systems and management productivity; necessary manpower; financial requirements; and anticipated sources of financing. The ORMVAs and specialized enterprises would sign maintenance contracts for three-year periods that would provide for preventive maintenance and major repairs of large or complex pumping stations.

----

1. The ORMVA Doukkala has already attained financial autonomy. Annual water charges (volumetric) vary from US $77 per hectare in the area irrigated by gravity to $230 per hectare in the area under sprinkler irrigation.

Cyprus: Pipe Conveyance System with Sprinkler and Drip Irrigation

The Southern Conveyor Project in Cyprus, which was financed by the World Bank, developed sprinkler and drip irrigation. The project had several objectives. The first was to secure a safe domestic water supply until at least the year 2000 for four areas that are major population centers--Nicosia, Limasol, Larnaca, and Famagusta--and the adjacent villages. About 75 percent of the population of the Greek part of the island is concentrated in these areas, and by the year 2010, about 83 percent of the population is expected to live in these areas--an increase from 435,000 residents in the 1980s to 589,000. The second objective was to provide irrigation water for maintenance of agricultural production in Kokkinokhoria, which has 9,125 hectares under cultivation, and four other areas along the south coast, where 1,330 hectares are currently under cultivation and irrigated agriculture is to be extended to 3,000 hectares gross. The pipe conveyance system will provide an incremental water supply of about 65 million cubic meters a year, of which 28.2 million would be utilized for domestic and industrial water supply and 36.8 million for irrigation. The project has several components:

Kouris dam, an earth-fill dam 103 meters high, provides a gross storage capacity of 115 million cubic meters.

Dhiarizos diversion consists of a headworks on the Dhiarizos river and a concrete-lined tunnel (diameter of 2.5 meters) 14.5 kilometers long.

The main conveyor, a ductile iron pipeline that is 110 kilometers long and that has a diameter ranging from 1400 to 800 millimeters, runs from Kouris reservoir to Akhna reservoir, with a connection to the Vasilikos-Pendaskinos project. Water will flow through the conveyor by gravity. The peak flow is about 3.8 cubic meters a second at the outlet of Kouris Dam and is reduced progressively to 0.5 cubic meters a second in its last section that supplies water to the Akhna reservoir. Break pressure tanks have been provided at four locations to prevent excessive static head and water hammer pressure.

Akhna terminal reservoir has a gross capacity of 5.8 million cubic meters behind an earth-fill dam 16 meters high. The reservoir balances the water supplies from the Kouris dam with the irrigation demand of Kokkinokhoria area. Its seasonal storage makes possible a nearly constant flow through the conveyor.

Facilities for the domestic water supply consists of the Limasol and Tersephanou water-treatment plant, the Tersephanou-Nicosia pipeline, and about twenty rural water supply schemes.

Pressurized irrigation distribution networks cover a total of 13,450 hectares.

A central control system has telemetric transmission for monitoring and control.

The five irrigation schemes will be connected to the main conveyor pipeline through one or more conveyor branches that supply night storage reservoirs, which will store the conveyor flow during the night and serve command areas of about 400 hectares. Flow through the branches might be up to 24 hours a day when by gravity and 20 hours a day when water is pumped; balancing reservoirs are provided where needed. From the night storage reservoirs, which are equipped with pumping stations where needed, pipe distribution networks, designed to deliver peak flows in 16 hours, will deliver water under pressure to hydrants at the head of the irrigation units. Pipes used in the network will be of asbestos cement for diameters of 150 millimeters or more and of PVC for pipes with smaller diameters. Each hydrant will have 1 to 4 outlets, each serving one irrigation unit. Each outlet will be provided with a filter, flow-limiting device, flow meter, and pressure regulator. Delivery to the farm plots--those not exceeding 2.5 hectares--at a maximum of three per irrigation unit will be through a farm outlet that is connected to the hydrant through a farm line of PVC pipe. Each farm outlet will be provided with a water meter; water will be delivered on demand. Land consolidation is planned for more than one-half the area. Total cost of the project (including Kouris Dam) in 1987 prices is US $330 million; with 50 percent of the cost allocated to irrigation, the cost per hectare is $12,300.

In the implementation of the project, high priority is being given to develop the irrigation system in the Kokkinokhoria area, which produces potatoes, a very valuable export for Cyprus. In the past, farmers have irrigated from ground water. The aquifer is in a dire situation, however, because of excessive exploitation since the 1960s, and it is endangered by intrusion of sea water. It is hoped that the aquifer can be restored with excess water from the pipe conveyance system and in the future can be operated in conjunction with surface water. The Akrotiri aquifer, by contrast, has been well managed and will continue to yield about 10 million cubic meters for irrigation.

The pipe conveyance system is expected to be the last large-scale water resource development project for Cyprus. An important water source to meet the needs beyond the year 2000 will be treated sewage water. Based on projected demand for water in Nicosia, Larnaca, and Limasol towns, estimates of potential sewage volumes are 38 million cubic meters a year by 2000 and 43 million cubic meters a year by 2010. The reuse of treated sewage for irrigation of agricultural areas would enable freshwater resources to be used for domestic consumption. Construction of the first sewage treatment plant near Limasol is planned, but because its location has become a sensitive political issue, its construction has been delayed. Under the pipe conveyance system, several feasibility studies and evaluations are planned. Under investigation will be the farmers' willingness to use treated sewage, their preference for the types of crops to be irrigated, the method of irrigation, the quality of treated sewage required, and the sites for storage of treated sewage. The feasibility of recharging the aquifers with treated sewage will be determined. The economic, social, political, and legal aspects of reusing treated sewage for both irrigation and aquifer recharge will be evaluated. The environmental impact of using treated sewage either for irrigating crops or recharging aquifers will be determined.

Strategies for the EMENA Region

A considerable part of the irrigation activities in the EMENA Region is expected to be rehabilitation and improvement of existing systems. Selection of appropriate technology for rehabilitated and new systems should take into account the relevant technical, socioeconomic, financial, and institutional factors. Appropriate technology varies from country to country, a factor that is especially important in rehabilitation projects. As the discussion of Egypt's irrigation systems illustrates, water resources planning and system operation should be important in developing a strategy for rehabilitation and improvement of irrigation systems. The use of treated sewage for irrigation will become more important in the EMENA region because of the increasing demand of domestic and industrial users for water. Water for irrigation will become increasingly scarce in several EMENA countries.

Chapter 4

TECHNOLOGICAL DEVELOPMENTS: THE ISRAELI EXPERIENCE

David Melamed

The area under irrigation since establishment of the state of Israel in 1948 has grown about eightfold--from approximately 30,000 hectares to 240,000 hectares--which represents around 50 percent of the total cultivated area of the country. Although the number of hectares is small in comparison with the area irrigated or irrigable in most other countries, it possible to derive certain valuable lessons from the course of irrigation development in Israel. This paper describes the process and draws conclusions based in part on experience with irrigation technology in both developing and developed countries.

Parallel to the expansion of irrigated areas in Israel, irrigation technologies underwent considerable development. Whereas in 1948 most of the area was irrigated by gravity systems, today there are few relics of these systems; almost all irrigation is pressurized, served by sprinkler, mini-sprinkler, drip, or mechanical move systems. The reasons for this transition are numerous, and if Israeli experience and technologies are to be transferred to other countries, whether developed countries or areas where traditional gravity technologies still predominate, some knowledge of the Israeli background is needed. No agricultural or irrigation technology can be transferred without proper analysis of conditions in the source and target areas. This is especially so when adapting technology used in the Israeli context. A set of rather singular, almost unique, conditions in Israel favored the widespread and rapid adoption of sprinkler irrigation technology at the end of the 1940s and in the early 1950s and the continuous innovation, whereby new technologies were adopted almost as fast as they moved out of the experimental stage.

Several features are characteristic of Israel's irrigated agricultural sector. Water resources are far short of meeting the demand for irrigation water for the country's irrigable area--around 1,200 million cubic meters for the present irrigated area of 240,000 hectares. Distribution of the scarce water resources in relation to land resources is poor; this factor, and the scarcity of water, prompted the development of a pressurized national water supply system supplying practically all areas of the country. Unit costs of water are high, which reflect the scarcity of water and the poor distribution of resources and the ramifications of a national supply system. Allocations of water to the irrigation sector and the individual producers are rigid; each producer is allocated a limited quantity for which charges are levied for the amount used. Irrigation labor costs are high, and most farmers want to carry out all work on the farm by themselves without using outside labor. Crop production systems are advanced and yields are high; hence water has a high marginal value.

Issues other than water are also significant in characterizing the Israeli setting. Farmers are relatively highly skilled; they are open to new technologies and motivated to increase their irrigation efficiencies. This cadre served as a catalyzing agent to the many new inexperienced farmers who settled mainly on smallholder cooperative farms. In addition, irrigation technicians specializing in irrigation can be found in all *kibbutzim* (large farms) as well as in many smallholder settlements. The irrigation industry is highly developed, and in many cases the firms are set up on *kibbutzim*. They are thus well aware of and tuned in to the needs of farmers. The firms are eager to experiment with and develop new systems and components designed to increase efficiency and cut irrigation labor inputs. In addition, the industry maintains intensive follow-up services for consumers.

A well-structured crop agricultural extension service exists, and one section of this service deals solely with irrigation and soils. The Irrigation and Soils Field Service has branches in all regions of the country, and in most areas it is staffed by farmers with academic qualifications. There are high-level research institutions that maintain a network of regional stations, which are closely involved and well-oriented to problems facing farmers and the agricultural sector as a whole. Government policy, which fostered investment in irrigation systems, is designed to improve water-use efficiencies and consequently the availability of low-cost capital.

These conditions encouraged the development of advanced irrigation technologies aimed at the attainment of maximum returns per unit of water through increased irrigation efficiency, high yields, and improved labor productivity. The scarcity of water motivated farmers and extension workers to make special efforts to increase irrigation efficiencies; efficiencies of 80 percent or more have been attained in most areas under sprinkler and drip irrigation. Partly due to this increase in irrigation efficiency (and partly because of extension and research on crop water use), annual crop water use has declined over the years rather significantly--from an average of around 800 millimeters to the present typical consumption in orchards (including citrus) of 700 millimeters, in truck crops of 450 millimeters per season, and in field and industrial crops of 500 millimeters per season.

Evolution of Irrigation Systems in Israel

Irrigation systems in Israel have undergone and continue to undergo continuous change. The massive transition dated back principally to the 1950s, when gravity irrigation systems gave way to pressurized movable sprinkler irrigation and a national and regional water conveyance infrastructure was constructed, which enabled sprinkler irrigation to be conducted in the hours of the day that are relatively free of wind. These developments greatly increased irrigation efficiency. Very few farmers still irrigate with gravity methods, although those that do appear to be quite happy with their systems, probably because of a successful combination of skilled irrigators and suitable local conditions.

Truck Crops

Movable sprinkler systems predominated throughout the 1950s and 1960s for irrigation of truck crops, and these systems are still in use in some truck-crop areas. The principal development in this period was the introduction of solid set systems; they were introduced to save labor and labor costs, which were particularly high because of the small applications and relatively high irrigation frequencies required.

In some crops, however, it soon became clear that irrigation that did not wet the foliage had distinct advantages, not only when using slightly brackish water but also when using high-quality water. Consequently, growers of certain crops susceptible to fungi attack (e.g., cucurbits) reverted to the use of gravity irrigation.

In the early 1960s, a number of field researchers began to try out various types of drip laterals, both below and above ground; at that time these methods were still in the experimental stage. They had two objectives in mind: to save water by reducing evaporation losses; and to see if drip irrigation could allow desert soils to be cultivated and to study the effect of brackish water on these soils. These trials showed that the above-ground system was promising from a technical aspect and that the desert areas previously considered almost useless for cropping could be irrigated successfully. Some saving of water was possible with drip lateral systems, but of greater importance were the considerably higher yields of truck crops that could be obtained than those attained with other irrigation systems, even in fertile soils. It was but a short step from this first experiment to the birth of a drip irrigation industry. Once industrialized, drip irrigation soon began to penetrate truck-crop areas, and this development soon caused growers to abandon gravity irrigation almost completely.

Once drip irrigation proved reliable, it quickly established itself as the main irrigation system for truck crops, especially for those crops grown under plastic, including those grown in tunnels or sleeves. It is also highly suitable for application of fertilizer through the system (fertigation). Yields obtained for several drip-irrigated truck crops are considerably higher than those obtained with other irrigation systems, not only in the problematic desert areas with brackish water but also in all other areas of the country.

Drip irrigation systems for truck crops are relatively expensive because of the close lateral spacing required, which is a function of the spacing between the crop rows. Some attempts were made to reduce costs by making systems partly portable, or not solid, by allocating one drip lateral for a number of rows and moving the drip lateral from row to row. But because of the high labor input required, this practice was soon abandoned. Most of the drip irrigated systems in truck crops are now solid and are laid out at the beginning of each season. At the same time, solid set mini-sprinkler systems with relatively close spacings have taken over some of the areas previously irrigated by solid set sprinkler systems.

The present situation in truck crops can be summed up as follows: practically all truck crops planted at relatively wide row spacings (80 centimeters or more), such as tomatoes and eggplant, are irrigated by drip systems, whereas most of the areas planted in closely spaced rows, such as carrots and lettuce (row spacings of 20 to 30 centimeters), are irrigated by mini-sprinklers.

Field Crops

In the late 1950s, farmers began to be weary of the hard unpleasant work involved in the hand moving of sprinkler laterals, especially in heavy soils and hot weather and in tall dense crops such as corn and cotton. The first step taken, aimed at reducing labor inputs, was to open up paths between the rows to facilitate moving the sprinkler laterals. Sprinklers with larger nozzle discharges and greater coverage were also introduced to allow wider lateral spacings, and as a result, sprinklers had to be moved less often.

This development was followed in the early 1960s by the introduction of longitudinal tractor-drawn end-tow laterals, which increased labor productivity and took much of the fatigue and unpleasantness out of sprinkler irrigation in these crops. End-tow systems also contributed to increased irrigation distribution uniformities and overall efficiencies, since entire blocks of eight or ten overlapping laterals were irrigated simultaneously, and so to some extent wind drift was offset. Costs of the system increased because pipes of larger diameters had to be used in the main system to convey the more highly concentrated discharges and supply the higher discharge required by the shorter irrigation rotation. The system also required additional accessories, and the equipment was subject to increased wear and tear.[1] These systems are still popular and account for some 50 percent of the field-crop area.

To overcome the difficulties encountered in irrigating dense field crops, some farmers reverted to surface irrigation, while others introduced closely-spaced under-tree orchard sprinklers mounted on skids or wheel carriages. The sprinklers were fed from flexible piping pulled by hand by the operator standing on the field border; sometimes stationary sprinkler risers were joined to the lateral by quick couplings, which allowed the sprinklers to be moved from one position to another. None of these alternatives proved viable, and they were abandoned within a relatively short period.

There are several reasons for the development of drip irrigation, which first appeared on the scene in truck crops and orchards in the mid-1960s. To solve the problem of irrigating field crops in marginal areas and in the windy hours of the day, farmers

---

1. Because of the rather limited field lengths (around 1,200 to 1,600 meters at most), these laterals that were generally 200 meters long could serve only six to eight positions, and consequently higher discharge concentrations were the result.

began to look to the first drip systems being introduced in orchards and truck crops at that time. Although drip systems at first seemed prohibitively costly, farmers nevertheless began to try these systems in marginal areas--particularly in odd-shaped areas not convenient for sprinkler systems, in areas of poor topography, and in shallow or stony soils--and in irrigated areas during the windy hours of the day. These trial areas responded well to drip irrigation, and in the more arid areas of the country, the incremental yields obtained under the drip systems appeared to justify the transition. Areas without problematic configurations, slopes, depth, or texture also had good results, and consequently there was a fairly massive transition to drip irrigation. Approximately 40 percent of the field-crop area in the country is now under drip irrigation. This transition was also fostered to some extent by the development of cheaper drip systems-- thin-wall piping and other cheaper components, which increased the feasibility of drip systems in field-crop areas.

But even with these reductions in system costs, drip irrigation entails a significant increase in costs at the plot level. The higher costs are offset in certain circumstances by the increased efficiency of the main system, since drip irrigation reduces the impact of wind as a limiting factor; theoretically, the systems can therefore be operated 24 hours a day no matter how windy, thus reducing hourly flows in the conveyance system and reducing pump loads. Although drip emitters require lower operating pressures than field-crop sprinklers, pressure requirements nevertheless remain more or less the same as those for sprinkler systems, since pressure losses in the system components, and especially in the filters, account for the difference between emitter and sprinkler pressure requirements. Moreover, the same system pressure is often needed, since certain crops require a portable back-up sprinkler system for seed germination and for activation of herbicides applied at the beginning of the season.

Fields in which drip irrigation have been introduced in Israel have in most cases been irrigated previously. The designs for drip systems had to be adapted to the existing permanent mains and secondaries, which were generally installed at 400-meter intervals in the large farms to serve the 200-meter long sprinkler laterals. But in the early 1970s, compensating pressure emitters were developed to permit drip irrigation of undulating lands, which ensured more or less equal emitter discharges irrespective of pressure head differences. Development of the pressure-compensating emitter made possible the use of 360-meter long drip laterals, with resultant economies in the secondaries. Along with the introduction of long drip laterals, tractor-drawn, power-operated reel systems were developed for mechanized laying out and rolling up of the drip laterals at the beginning and end of the season; these systems helped reduce labor costs.

The main problem encountered in drip irrigation--that of emitter clogging--has been largely overcome by the development of the labyrinth emitter, which was used in place of the original laminar-type emitter (the labyrinth emitter also facilitates use of longer drip laterals) and the development of improved filters, including the self-flushing filter. Drip irrigation from open reservoirs that store marginal water is thus no longer a problem, and large areas of field crops, particularly cotton, are drip-irrigated from open reservoirs. Even so, blockages do occur with some water sources, and these problems call for special investigation and pretreatment methods.

In the last few years, mechanical move laterals and irrigation booms, which require little labor, have been introduced successfully, and these systems are gaining popularity. Their main limitation is their relatively high rate of instantaneous precipitation that frequently results in runoff in fine-textured soils.

Orchards: Deciduous, Citrus, and Sub-Tropical Fruits

Movable sprinkler systems (under-tree sprinklers) were until recently the system used most often in orchards, although basin irrigation was common in certain areas of the country. To cut the labor costs entailed by these systems, farmers had a choice of three systems: solid set overhead sprinkler systems generally with 18- and 24-meter sprinkler lateral spacings; tractor-drawn longitudinal end-tow lateral systems; and under-tree sprinklers mounted on hand-dragged plastic laterals mounted on skids.

But in the mid-1960s drip and micro-sprinklers began to penetrate, fostered particularly by the development of a local polyethylene pipe industry. Practically all the orchards are now irrigated by drip and micro-sprinkler systems. Farmers use micro-sprinklers of different capacities to increase the wetted area as the root system expands and the orchard matures to full production stage. But approximately 10 percent of the orchard area is still irrigated by sprinklers mounted on hand-drawn plastic laterals.

Development of Automation

For several reasons, much attention has been given in Israel to the development of automated irrigation and, specifically, to automation of various system components. The first step to automation was the development of the metering valve. This type of valve, which shuts off the flow as soon as a certain quantity has been delivered, is not affected by the pressure fluctuations that occur when an automatic valve is time-controlled. Introduction of metering valves has made a major contribution to increased irrigation efficiency.

This development was followed by sequential automation devices (also based on metering valves), which switched the irrigation to the second, third, or fourth block and so on, in sequence, after the set application has been delivered to the first block. In addition to increasing irrigation efficiency, this level of automation also reduced labor inputs and facilitated continuous operation of the system and fuller exploitation of the night hours.

More recently, central computerized command systems have been introduced, and irrigation of entire farms can be centrally controlled from one command point. Auxiliary system components, such as self-flushing filters, are also now automated. Another outcome of this development has been the introduction of proportional fertilizer systems, which apply fertilizer during an irrigation period (such as mid-third) and allow fertilizer concentrations to be predetermined. Other developments include software programs that can shut down the system or part of it if, for instance, wind speeds begin to exceed the permissible speed prescribed for the sprinklers or if system failures, such as a break in the pipe system, cause water losses or pressure drops. These programs also enable the flow to be transferred to areas irrigated by under-tree sprinklers, such as orchards, which are unaffected by high wind speeds.

Adoption of New Technology

Transfer of an irrigation technology, know-how, and experience from one area to another should be made only after in-depth analysis of agricultural and irrigation parameters show that conditions in the target area are sufficiently similar. It is generally recommended that supporting services, such as agricultural and irrigation extension, research, and technical back-up services should be built up in the target area in tandem with the transfer of the new technology.

The main problem encountered in selecting an irrigation technology is the difficulty of quantifying the large number of irrigation-related factors. Consideration of a new technology should be done side-by-side with a careful, unprejudiced evaluation of the existing technology. This point is particularly important. At times one can get hooked to a new technology, and the enthusiasm to innovate must then be tempered by the realization that a new technology often has its own dynamics. One can easily be blinded to the fact that in a particular development context there might not be any viable alternative to the traditional technology in use. In such cases, the effort to improve irrigation efficiency must therefore rely primarily on extension services.

Evaluation of physical constraints and other factors immediately reduce the technological options and eliminate those technologies that are clearly unsuitable. The objectives aimed at and the expectations sought from the remaining technologies must then be defined, and the prospects for realizing these expectations from alternative technologies must be estimated.

Expected Benefits from New Technology

Typical expectations from adoption of a new irrigation technology include the following: increased farm profits; increased production per unit of water; reduced irrigation labor inputs; increased employment opportunities as a consequence of more intensified cropping or expansion of irrigated area; and efficient capital utilization. These expectations can be realized by attaining higher irrigation efficiencies, greater land utilization, and higher yields and by reducing labor inputs.

Irrigation Efficiencies. If the first of these expectations, higher irrigation efficiencies, can be attained, increased income will result as a consequence of increasing the irrigated area from a limited water resource, avoiding damage from waterlogging, reducing the costs of subsurface drainage systems, reducing energy costs, and reducing the dimensions and costs of water conveyance systems. The effect of increasing irrigation efficiency should generally be studied at the overall project or area level, since it may be feasible to irrigate at low water use efficiencies at the farm level (provided a minimum distribution uniformity is obtained) and to return runoff or deep percolation via regional collectors to the main conveyance system and thus still attain an overall high project-level water-use efficiency.

Land Utilization. The second objective is greater land utilization. Land utilization ratios are primarily a function of cropping patterns, but under certain technologies, particularly the type of water conveyance system used, not all the land is utilized. Unlike pipe conveyance systems, irrigation ditches reduce the net cultivated area; under certain conditions, yields in areas adjacent to the ditches may be affected.

With the standard type of center-pivot system, part of the area is left unirrigated. With sprinkler systems, the crop on the plot perimeter may be inadequately irrigated, and the land is thus not properly utilized. Part-circle sprinklers can be used, however, and the lateral can be placed on the perimeter, a practice that is practical only with solid set systems. Water is wasted if rectangular sprinkler laterals are placed on the perimeter; this can be a major consideration in small plots.

In contrast to other methods, drip systems fully exploit the entire area of the plot, including marginal areas with such characteristics as shallow soils, steep topography, awkward shape, and so on. This factor is of major importance if arable land is scarce, but it is also of economic significance in all irrigation projects, since it has a direct influence on specific irrigation costs and general area infrastructure costs per unit of cultivated area. On-farm irrigation system costs in Israel at the plot level, exclusive of costs of the main system, range from around $1,000 to $3,200 per hectare (with the exception of solid set systems in truck crops), according to the crop and technology used (see Table 4.1).

Higher Yields. Farmers expect higher yields to result from the adoption of new irrigation technologies because they can have improved control over several important

Table 4.1

On-farm Irrigation System Costs, Israel

Crop Irrigation System	Cost (US dollars per hectare)

Truck Crops

Hand-move sprinkler laterals	1,400
Drip-solid set	3,000
Mini-sprinklers - solid set	3,100
Solid set sprinkler laterals	5,700

Field Crops

Hand-move sprinkler laterals	1,000
End-tow laterals	1,600
Mechanical move laterals	1,600
Drip - seasonally solid	2,500
Drip - seasonally solid - thin wall	1,300

Orchards

Hand-move sprinkler laterals	1,600
Sprinklers on plastic drag lines	2,000
Overtree sprinklers - solid set	3,200
Drip - solid set	1,500
Micro-sprinklers - solid set	2,200

Grapevines

| drip | 2,200 |

factors. They are able to apply appropriate quantities of water at the required frequency. They can avoid wetting the plant foliage and can thus reduce the incidence of plant disease and leaf burn from salts in the water. They are able to combine various agrotechnical treatments, such as fertilizer applications and solar treatment to control weed infestations and soil disease. Waterlogging can be avoided, and salt accumulation can be minimized.

Strangely enough, the likelihood of obtaining higher yields as a consequence of introducing a new technology is not generally taken into account by financing agencies. Yields are generally taken as constant, and the comparison of different technologies is based only on expected irrigation efficiencies, that is, the quantity of water required to obtain the potential yield. This approach often negates the introduction of a new technology.

There is no guarantee, however, that adoption of a new technology will indeed give the expected increases in yield. Deviations from expectations might prove significant. It appears that the greater part of the potential yield can be obtained by practically all available systems, provided they are properly designed, constructed, and operated. In certain circumstances, however, efficiently designed, installed, and operated systems are not capable of obtaining optimal yields. For crops sensitive to wetting, for example, optimal yields cannot be obtained. Very high increases in yields obtained following adoption of a new irrigation technology are often due to the simultaneous injection of other technological changes and agrotechnical improvements with the introduction of the new irrigation technology. Alternatively, previous operation of the system might have been suboptimal.

Reduced Labor Inputs. The expectation of lower irrigation labor inputs acts as an incentive for introducing new irrigation technology, but it is an expectation that varies with the level of economic development. New technologies are adopted primarily in advanced agricultural societies in which labor costs are high. In developing countries, labor costs are low, and government policy often aims at generating jobs rather than expending capital to reduce labor.

Implementation Issues

Each type of system has its own potential for irrigation efficiency. But realization of this potential is sensitive to the level of performance in each of the three main phases of project implementation: planning and design, construction or field installation, and day-to-day operation. Subject to this sensitivity, actual irrigation efficiency will deviate from the expected or target efficiency. Estimating the range of efficiencies, or the average deviation from the potential, is probably more important or at least more practical than considering the absolute efficiency potential of a particular technology. This is especially important when a large-scale development project is being planned and the question of how the individual system operators at the farm level will manage their systems has to be considered.

In the planning and design phase, greater expertise and experience, as well as more detailed data, are required in the planning of gravity systems than in other systems. Information on soil conditions, especially infiltration properties, and detailed, accurate topographical mapping are needed.

In the construction or installation stage, major difficulties are encountered in carrying out accurate land forming works. Even if suitable earthwork contractors are available, difficulties can arise later in system maintenance and periodic rehabilitation of the works. Supervision of the land forming works is also problematic. In evaluating system expectations, planners must give due consideration to this phase of the work. It might be advisable to disregard an otherwise suitable technology, such as minimum slope furrows, because there is no guarantee that the land forming work will be sufficiently accurate. These systems would thus not provide the advantages expected.

In day-to-day operation, sensitivity to operational expertise varies with the different types of systems. Contrary to the widely held view, gravity irrigation systems, in which the greater part of the infiltration to the soil occurs while the system is flowing along the furrow or border, requires the greatest expertise to obtain high efficiencies. The irrigator must know how to regulate the water level in the ditch, estimate and regulate stream flows, calculate the required duration of stream flow in the furrow for wetting the root zone to the required depth, and estimate distribution uniformity along the run in order to make adjustments in later irrigations. Moreover, successful operation is also dependent on the correct estimation of the variation in soil conditions (its infiltration capacity) during the course of the season. Stream flows have to be varied (initial wetting and cut-back stream) during the course of each irrigation, and this fine tuning also requires experience.

Of all the gravity systems, the simplest to operate are the dead-level border and furrow systems in which most of the infiltration into the soil results from ponding. If the stream size is known and it remains constant and if it is sufficiently large to wet the run within the specified design time, all the operator has to do is to determine the required flow duration and shut off the flow accordingly.

Sprinkler irrigation systems, especially solid set systems controlled by metering valves, are the simplest to operate efficiently. The main parameters, such as nozzle diameters and discharges, sprinkler spacings, and precipitation rates, are predetermined, leaving only one easily determined variable, the time or duration of irrigation, for the operator. If metering valves are used, water quantities are measured directly, and the laterals shut off automatically without the operator having to take into account any pressure fluctuations that occur during the course of irrigation. Assuming pressure and wind conditions are satisfactory, all the operator must do is to keep an eye open for nozzle blockages--so long as appropriate filtration is built into the system and is maintained.

Efficient operation of drip systems is also easily attained, although not as easily as with sprinkler systems. More maintenance--such as flushing of drip laterals and chemical applications to remove emitter clogging--is required. In both sprinkler and drip systems, irrigation is also readily managed, and the instructions of crop and irrigation extensionists are easily translated into field practice. All the systems discussed above require a greater or lesser degree of mechanical maintenance services. Mechanical move and center-pivot systems are most in need of these services, but technical back-up services are also needed for other pressurized systems and for systems in which regulating and automating devices are incorporated.

Suitability of Irrigation Systems to Specific Conditions

Furrow irrigation with the furrows aligned with the dominant slope and with the furrow stream diverted by siphons is best suited to areas with uniform slopes of 0.2 to 0.3 percent. In these conditions, land-leveling work is the minimum required in any case to ensure adequate surface drainage. Contour head ditches, with practically no structures, are easily operated and can generally be excavated with mechanical equipment available on the farm. These ditches require no embankments, can be obliterated at the end of a season, and require practically no maintenance.

The main and most outstanding advantage of this system is its very low construction cost, which in many cases is a most important consideration. It can be implemented in stages, with land leveling in the first stage designed to give only a continuous slope. The area can be brought under irrigation quickly, thus increasing project feasibility. In the second stage, when water becomes a limiting factor, land-forming operations can be carried out to obtain a uniform slope. The irrigation efficiency to be expected from this system is not high, however.

With the use of contour furrows and borders, good results can be obtained, but construction of the system calls for highly accurate implementation of earthworks and careful inspection and supervision of the work carried out by contractors. If there is little experience with construction and operation of these systems in the area, it is advised to adopt other systems, unless it is possible to import expertise, such as laser-controlled leveling, from other areas.

High efficiencies can be obtained with sprinkler irrigation and micro-sprinklers if wind velocities during most of the day are not excessive. The system is easy to construct and its operation simple. Where labor is not in short supply or costly, hand-move systems can be selected, in which case the systems are relatively cheap. Metering valves and other automation systems can increase efficiencies still further. Solid set micro-sprinklers are recommended for orchards because they are the least costly solid sprinkler irrigation system, and they have advantages in efficiency of water use.

Drip irrigation systems give high efficiencies and in some crops higher yields than those obtained with other systems. Small frequent applications of water can be given without difficulty, water is continuously available, the plant foliage is not wetted, and fertilizer applications can easily be integrated with irrigation.

The main drawback in drip irrigation is the sensitivity of the emitter to clogging, especially when surface water sources are used. Drip irrigation systems need to be introduced gradually, particularly if the water is from a surface source, and only after information on water quality has been gathered and assessed. No reliable criteria have yet been formulated for classifying water according to its potential for clogging.

There are other drawbacks to drip irrigation. Its costs are relatively high, especially for closely-spaced truck crops. But when the crop rows are widely spaced, as in orchards, solid drip systems are no more expensive than hand-move sprinkler laterals. Salinity is a hazard. Under certain conditions, salts accumulate on the perimeter of the wetted "bulb," and rain may cause these salts to leach into the root zone. In many cases, a complementary portable sprinkler system may be required at the beginning of the crop season to ensure proper germination and to activate herbicides. Either prior to seedbed preparation or after the growing season, a complementary sprinkler system may be required for leaching of salts.

Suggested Irrigation Technologies and Transition Processes

A transition to more advanced irrigation technologies can be considered in certain settings: capital resources are limited; labor is not a limiting factor, and labor costs are low; farms are small, and most of the farm operations are carried out by the farm family; and supporting and technical services are still in the early stages of development.

In ideal conditions for field and truck crops, where natural slopes are less than 0.2 percent and infiltration is satisfactory, furrow irrigation is suggested, with furrows aligned with the dominant slope if water is plentiful, water use efficiency is not critical, and natural drainage is satisfactory. If water economy is important, water-use efficiency can be improved by construction of return flow systems. In areas of steeper or irregular slopes, hand-move sprinkler lateral systems are suggested if wind speeds are not excessive. If the level of crop production and intensification justify the investment, further upgrading of irrigation could be by staged introduction of drip systems, or if soil and topographical conditions permit, laser-controlled land-forming operations can obtain uniform low-grade slopes. If sprinkler and drip systems are to be introduced, metering valves should form an integral part of the system; they provide the simplest method for controlling water quantities.

If soil and topographical conditions are ideal and water use efficiencies are not critical, furrow and border systems may be used for orchard crops. The general trend, however, should be toward introduction of hand-move, under-tree sprinkler systems with metering valves. Subsequently, permanent fixed systems can be installed, but drip or micro-sprinkler systems (the choice made according to orchard crop) are suggested for attaining the highest efficiency and for the best yields. Drip irrigation is recommended for grapevines.

Chapter 5

TECHNOLOGICAL DEVELOPMENTS IN THE UNITED STATES

Charles Burt

Agricultural irrigation systems in California, which cover over 10 million acres, are highly diverse. Virtually every viable method of irrigation used in the world can be found in California. Of the total acreage, 76 percent is covered by surface irrigation methods--furrow, border strip, rice, and basin--and 24 percent is irrigated by pressurized methods, either drip or sprinkler. This paper focuses on technological developments in California irrigated agriculture and efforts to increase irrigation efficiency and thereby reduce energy and water consumption.

Factors in Irrigation Efficiency

A study in 1976 by the U. S. General Accounting Office found that the average on-farm irrigation efficiency in the United States is 50 percent.[1] At first glance, this would appear to mean that twice as many acres could be farmed if the irrigation efficiency was improved to 100 percent. Studies in California, however, have indicated that achievable water savings in California due to agricultural water conservation may be only about 0.8 million acre-feet (3 percent of the 30 million acre-feet used annually), which represents an irrigation efficiency above 90 percent. Conflicting reports such as these are the norm rather than the exception in discussions regarding irrigation efficiency. The difference often lies in the definition of "irrigation efficiency." Low on-farm irrigation efficiencies do not generally result in a loss of water to a basin or ground water hydrological unit. The water that runs off or deep percolates from one field or farm is generally used by another farmer downstream. It is entirely possible, therefore, to have low on-farm irrigation efficiencies, as reported in the GAO study, and at the same time to have high basin irrigation efficiencies.[2] In this discussion, the "irrigation efficiency" that is described refers to "on-farm irrigation efficiency" unless otherwise noted. Irrigation efficiency (IE) is defined as:

----

1. See the study by the General Accounting Office (GAO), Better Federal Coordination Needed to Promote More Efficient Farm Irrigation. Report to Congress. (Washington, D.C.: U. S. Government Printing Office, June 1976).

2. See R. M. Hagan and D. C. Davenport, "Agricultural Water Conservation in Simplified Perspective," California Agriculture (November-December, 1981).

$$IE = \frac{\text{Water Beneficially Used}}{\text{Water Applied}} \times 100$$

Water applied by irrigation has many possible destinations. The water may go into the air as evaporation, spray loss, or transpiration, be stored in the soil root zone, run off the surface of a field, or exit the root zone as deep percolation. This deep percolation can either contribute to high water tables or can recharge streams and ground water aquifers, or both. The deep percolation can be either beneficial or harmful to neighboring farms. Table 5.1 defines the major forms of "beneficial," as opposed to "non-beneficial," water uses in terms of on-farm irrigation efficiency. There is ample justification for improving low on-farm irrigation efficiencies, for they can result in the following: excess pumping, fertilizer leaching, low crop yields, degradation of water quality, drainage problems, excessive water costs, and reduction in the acreage that can be irrigated with a fixed volume of water available on a farm.

Energy Use and Irrigation Efficiency

Agricultural irrigation plays a major role in water and power consumption in California. The 30 million acre-feet of water used per year for agricultural irrigation represents 85 percent of the diverted or pumped water in the state.[3] Electricity powers over 90 percent of the agricultural pumps in use at the farm level in California.[4] Approximately 5 billion kilowatt hours of electricity are used annually for pumping.[5] These figures do not include the pumping requirements of the many irrigation districts or Federal and State canal systems, such as the California Aqueduct.

An annual power bill is computed according to the following formula:

Annual energy cost =

(Cost in dollars per kilowatt-hour)*(Kilowatts)*(Hours of pumping per year)

----

3. See R. M. Hagan and D. C. Davenport, "Agricultural Water Conservation in Simplified Perspective," California Agriculture (November-December, 1981).

4. See California Department of Water Resources, Water Conservation in California DWR Bulletin 198-84, 1984.

5. See G. Kah, D. Whitson, and M. Seedorf, "Agricultural Energy M██████ement in California: A Utility's Perspective," ASAE Paper no. 83-3520, 1983.

Table 5.1 Beneficial and Non-beneficial Uses of Water for On-farm Irrigation

Beneficial Uses	Non-Beneficial Uses
Crop transpiration	Weed transpiration
Leaching for salinity control	Deep percolation in excess of leaching requirement
Special practices	Evaporation from wet soil surfaces
-packing the soil for harvest	Evaporation from wet foliage
-weed germination	Canal and pipe seepage
-climate control	Spray losses

where

the cost per kilowatt hour is the average electricity charge,

kilowatts are determined by motor size (1 Hp = 0.746 Kw), and

hours of pumping per year depend on the irrigation scheduling.

The horsepower (Hp) is calculated as follows:

$$Hp = \frac{GPM*(Feet\ of\ Pressure)}{3960*(Pump\ Efficiency/100)*(Motor\ Efficiency/100)} \quad ,$$

where

GPM is the pump flow rate in gallons per minute, and

Feet of Pressure is the total pumping pressure.

The on-farm irrigation efficiency directly affects the horsepower in two ways: a low irrigation efficiency makes it necessary to have a large pump flow rate; and a system with poor irrigation scheduling will be operated too many hours per year. The factors affecting annual energy use for irrigation pumping fall into the following categories:

Irrigation Efficiency

Irrigation timing (scheduling)

Uniformity of water distribution throughout a field

Unrecovered losses (evaporation, uncollected runoff)

Other

Pump efficiency

Motor efficiency

Pressure requirements

- Depth to water

- Drawdown in a well

- Pipe, valve, and column friction

- Final discharge pressure (for example, the sprinkler pressure)

- Elevation change

Spray and Evaporation Losses

Under normal irrigation practices, spray and evaporation losses constitute a rather small percentage of the total water applied. Typical values are given in Table 5.2. Instances in which these losses can be appreciably higher are associated with sprinkler irrigation. Hot and dry winds, high sprinkler pressures, and short set durations can combine to give evaporation losses greater than 60 percent.

Uniformity of Water Distribution

The "distribution uniformity" describes how evenly water is made available to plants throughout a field. Distribution uniformity (DU) is defined as follows:

$$DU = \frac{\text{Minimum depth infiltrated} \times 100}{\text{Average depth infiltrated}}$$

where the "minimum" is average depth infiltrated in the region receiving the lowest 25 percent of the water.

Ideally, irrigation methods should apply water uniformly to each plant in a field. Such a method would have a distribution uniformity of 100 percent. In fact, no system is capable of applying water so that every plant in a field receives the same amount of water. For all methods, some points in the field are always either over- or underirrigated, or both. A low degree of uniformity results in a waste of energy because excess water must be pumped onto the field to apply enough water at the dry points.

Figure 5.1 shows the relationship between the depth of water infiltrated and the portion of field area receiving that depth for an assumed surface irrigation water distribution pattern. The solid curved line indicates the distribution of water for an average of 4.0 units of water infiltrated. The average depth in the low quarter is 3.6 units. The distribution uniformity for this example is (100 x 3.6/4) = 90 percent. If the soil moisture deficiency (SMD) on the field just prior to irrigation is equal to 3.6 units, that area of the field to the right of point A in Figure 5.1 will be underirrigated, and the dotted area to the left of point A represents water lost by deep percolation. Adequate irrigation of the entire field area would require an average application depth of about 4.25 units.

The economics of irrigation system design may dictate that less than 100 percent of the area be adequately irrigated. Where the average of the low quarter is equal to the desired depth of application, approximately 12 percent of an irrigated field will be underirrigated. Figure 5.2 illustrates a simplified infiltration pattern with two distribution uniformities; the same average depth is infiltrated in both cases.

Table 5.2 Typical Annual Spray and Evaporation (Wet Soil and Plant Surface) Losses for Various Irrigation Methods

Irrigation Method	Spray and Evaporation Losses (percent)
Hand-move sprinkler	10 - 20
Undertree sprinkler	5 - 8
Surface	3 - 5

Figure 5.1 Example of Distribution Uniformity of Water Infiltrated along a Furrow

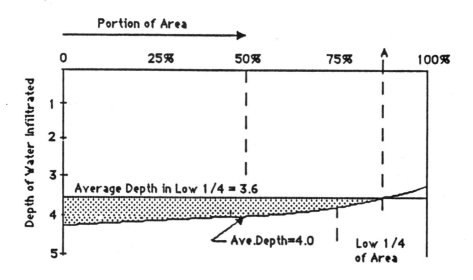

Figure 5.2 Simplified Infiltration Pattern with Two Distribution Uniformities

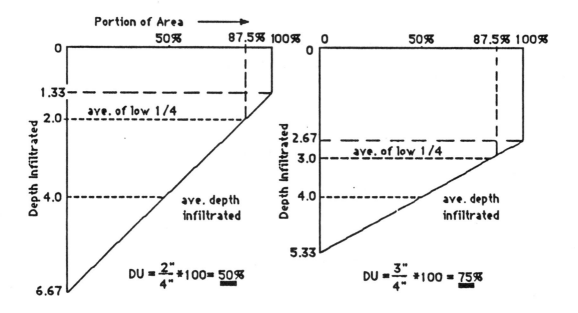

The distribution uniformity is not a measure of efficiency, because it does not quantify beneficial use or even deal with non-infiltrated water. A high irrigation efficiency with a fully irrigated field is only possible if there is a high distribution uniformity, however.

Various irrigation specialists put slightly different numbers in tables such as Table 5.3. The important point is that actual distribution uniformities are lower than potential values. Uniformities under field conditions depend upon three factors. The first of these is irrigation system design. Systems must be designed and installed properly. The second is suitability. Each system has a particular set of conditions under which it will operate satisfactorily. For example, furrows are not suited for sandy, non-homogeneous soil. Drip irrigation is not suitable for areas with large rodent populations because the rodents eat the hose. The third factor is management. Pressure regulators may need to be adjusted properly on pressurized systems. Large furrow flow rates are needed to attain a good advance ratio. There are generally many factors that affect the uniformity of water distribution to plants within a whole field. Table 5.4 lists various components that contribute to the distribution uniformity for an irrigation system; it can be seen that the uniformity of a hand-move sprinkler system is much greater than "catch can uniformity" and that the uniformity possible with sloping furrows consists of more than opportunity time differences alone.

Irrigation Efficiency

Figure 5.3 illustrates a case of perfect irrigation timing in which applied water has the following destinations: evaporation losses; runoff (which may or may not be lost); root zone storage; and deep percolation, below the root zone. All of the field is adequately irrigated. In the figure, the "stored" and "deep percolation" water represents water that infiltrated into the soil. All deep percolation is due to nonuniformity. The figure illustrates the fact that if no spot in a field is underirrigated, then all of the field is overirrigated.

Figure 5.4 illustrates a situation of overirrigation at the driest point due to improper timing. The sprinkler, drip, or furrow set length was too long. Deep percolation is due to both poor timing and nonuniformity.

Figure 5.5 shows a different concept. Some farmers believe that they have no deep percolation losses if the average depth infiltrated equals the soil moisture deficit. As can be seen, this would result in *both* deep percolation and underirrigation.

Figure 5.6 illustrates a situation that may be possible only with a dry soil surface and a buried drip system after the crop has been germinated by another system. There is no runoff or evaporation loss from the dry soil surface at this time. In this case, a 100 percent irrigation efficiency can be achieved by underirrigating the complete field (except one point). All the water will be used by the crop. The problem with this

Table 5.3 Potential Distribution Uniformities
for Moderately Well Designed Irrigation Systems

Irrigation Method	Potential Distribution Uniformity[a] (percent)
Permanent undertree sprinkler	94
Linear move	93
Sloping furrow	91
Orchard drip	90
Level furrow	87
Border strip	85
Row crop drip	80
Hand move sprinkler	75

Note: Irrigation methods are listed in order of decreasing distribution uniformities.

[a](Minimum depth/average depth)*100.

Table 5.4 Factors in Distribution Uniformity for Three Irrigation Methods

==

Method	DU Components	Factors within each component
Drip	-Flow rate variation	Pressure differences down a hose.
		Pressure differences between hoses.
		Plugging.
		Manufacturing variation.
		Leaks.
		Emitter part wear.
	-Unequal drainage	Low spots may drain for hours.
	-Variable tree and emitter spacing	Number of emitters/tree in various blocks of trees.
Hand Move sprinkler	-Catch can unif.	Nozzle pressure.
		Sprinkler spacing.
		Wind, angle of nozzle.
	-Flow rate diff.	Pressure variations along lateral.
		Pressure variations between laterals.
		Nozzle plugging.
		Nozzle wear (sand).
		Different nozzle sizes.
	-Leaks	
	-Unequal drainage	
	-Edge of field effects	
Sloping furrow	-Opportunity time differences	Advance time ratio for a single furrow.
		-flow rate
		-furrow length
		-soil intake characteristics
		-slope
		-smd
		Set time differences.
		Uneven land grading.
	-Soil variation	
	-Flow rate variation	Different flow rates to various sets

==

Figure 5.3 Destination of Applied Water for a Case of Perfect Irrigation Timing

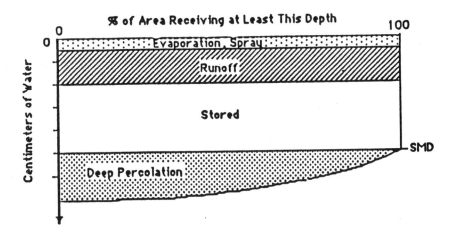

Figure 5.4 Destination of Applied Water for a Case of Overirrigation

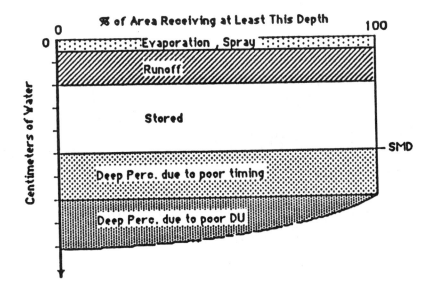

Figure 5.5 Destination of Applied Water when Average Infiltration Water Depth Equals the Soil Moisture Deficit

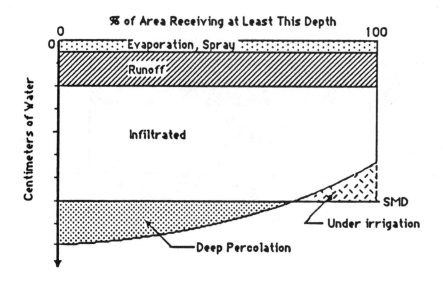

Figure 5.6 Destination of Applied Water for a Case of 100 Percent Irrigation Efficiency

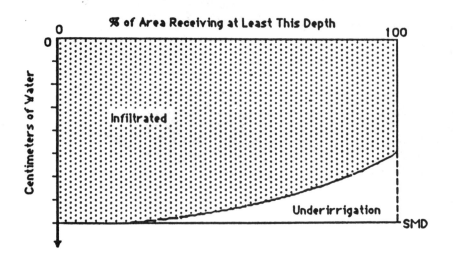

underirrigation approach is that the yields are unacceptably low, and salinity problems would develop.

Fertilizer Leaching. Water that percolates below the root zone (leaches) carries salt with it. Some leaching is necessary to remove salts that are brought in by irrigation water itself. The excessive leaching caused by nonuniformity and poor irrigation scheduling, however, results in heavy fertilizer losses. This fertilizer later shows up in the ground water, and this situation is beginning to cause serious concerns throughout the United States.

Nitrogen fertilizers are particularly susceptible to leaching loss. In addition to the aspect of ground-water pollution, this is an important energy conservation concern. One ton of nitrogen fertilizer requires approximately 42 million BTUs to manufacture. It has been estimated that only 54 percent of the applied nitrogen in the San Joaquin Valley in California is used by the plants. The remainder is either lost in leaching or volatilization.

Runoff. Runoff (tailwater) with surface irrigation systems does not decrease efficiencies if it is collected for reuse later somewhere on the farm. Runoff does not have to be recycled on the same field or irrigation set in order to qualify as "beneficial use." Runoff (which is collected) is a sign of good management in arid areas, because the existence of tailwater indicates good advance ratio and good distribution uniformity.

Obtaining a Good Irrigation Efficiency. In summary, a good irrigation efficiency can be obtained by achieving a good distribution uniformity, proper scheduling of irrigations in terms of time and quantity, collection of runoff, and minimal evaporation and spray losses.[6] Some of the other aspects of efficiency include return on investment, yield per unit of water applied, adaptability to existing social patterns, dependability of operation, and so on. These factors are generally much more important to the success of a project than a simple measure of irrigation efficiency as defined earlier. The following quotation attempts to address the difficulty of realistically assessing the "efficiency" of irrigation:

> Irrigation engineers are naturally concerned with the efficient use of water and financial resources. The irrigation process, however, involves the use of many other resources. We should begin to study the efficiency with which we use these resources as well. The task of developing significant and meaningful efficiency measures is not an easy one: different situations, studied from different points of view, require different efficiency measures. ... We irrigation engineers have long thought of ourselves as water managers. We must begin thinking of ourselves as resource managers.[7]

----

6. Except for some moving systems such as linear moves, or fast cycling systems such as solid set sprinklers, evaporation and spray losses may be a minor concern relative to other factors.

7. Kenneth H. Solomon, "Resource Utilization Efficiency," Transactions of the American Society of Agricultural Engineers, vol. 25, no. 1 (1982), pp. 93-95.

On some irrigation projects, the limiting factor may be something as simple as the ability to obtain spare parts for the system. Table 5.5 addresses such factors affecting irrigation efficiency. It can be seen that all irrigation systems have merits and disadvantages, and special designs and considerations always provide exceptions to the rule. Certainly there is no such thing as a perfect irrigation system. For each situation, the proper irrigation system should be selected after an investigation of the factors affecting efficiency, cost, and manageability.

Evaluating Irrigation Systems for Water and Energy Conservation

On-farm irrigation system evaluations are performed by a variety of agencies, both public and private, for a wide range of purposes. For example, agricultural energy management specialists from Pacific Gas and Electric Company (PG&E), California's main utility, conduct a rapid survey of a customer's irrigation system. The objective is to identify the potential for energy savings. Aspects addressed are irrigation scheduling, distribution uniformity, and pressure losses throughout the system.

The California Department of Water Resources, Office of Water Conservation Mobile Lab Teams maintained seven labs, composed of one irrigation specialist and several students each, that operated in California during the summer of 1986. Working through irrigation districts or resource conservation districts, they attempted to evaluate annual irrigation efficiency, distribution uniformity, annual power and water savings, and management constraints that could be corrected. A group of "expert system" computer programs were developed to assist in making these evaluations.[8]

In the Westland Water District of California, private consultants were brought in. The water district subsidized on-farm irrigation evaluations in the 42,000-acre area experiencing a high water table and underlain by drains supplying Kesterson Reservoir. The evaluations were performed by private consultants, and the objective was to reduce the drainage outflow, thereby reducing the selenium contamination problem at Kesterson.

When conducted by competent, well-trained irrigation specialists with a clear goal in mind, rapid evaluations can effectively pinpoint water and energy savings, which benefit individual farmers and society as a whole through reduced energy consumption, fertilizer leaching, and so on. Because of the tremendous complexity of irrigation systems and cropping patterns in California, it is much more difficult to evaluate irrigation systems in California than in most other areas of the United States.

----

8. See C. M. Burt, R. W. Walker, and S. W. Styles, "Irrigation System Evaluation, 1985" (Department of Agricultural Engineering, California Polytechnic State University, San Luis Obispo, California, 1985).

Table 5.5 Comparison of Irrigation Methods and Factors that Affect Irrigation, Power, and Operating Efficiency

ITEM	Level furrow	Sloping furrow w/ return flow	IRRIGATION METHOD — Sprinkler — Linear move	Center Pivot	Hand Move	Drip
Dirt in water	No effect	No effect	Sand separator and/or filter needed. Extra system pressure necessary= 3-10 psi			Extremely sensitive Filtration + chemical inject. needed
Maximum slope	Level	up to 3%	4%	15%	15%	Anything farmable
Min. economical field size, hectare	No minimum	4	100	50	1	2
Field shape	Any	Any	Long and Rectangular	Circular or modified circle	Any, but major edge effects on odd shapes	Any
Energy for install. and materials	Medium	Mod. low	Mod. high	Medium	Mod. high	Mod. high
Annual pumping $	None-V. low	Low	Medium	Mod. high	High	Mod. high
Technical expertise needed by farmer	Medium	Medium	Very high	High	Medium	Extremely high
Relative # of moving or complex parts (1=few, 10=many)	1	2	6	5	3	10
High salt water	Special bed shapes required, highest sensitivity during germination and seedling stages		Toxicity and leaf burn after plant has leaves. Good for germination		More serious problems than linear move or pivots	Best, but soil needs leaching by other means every five years
Non-uniform soils	-----Sensitive-------		------------------------Non-sensitive----------------------			
Inflexible water delivery system	Moderately ------tolerant--------		Very sensitive	Sensitive	Moderately tolerant	Sensitive
Minimum SMD, cm.	4-7	2.5-7	0.3	0.3	2.5	0.3
Fertigation ease	Good	Good	--------Excellent---------		Good	Excellent
Potential IE w/ Xlnt. mgmt. & design	88	88	90	85	75	88
Irrig. Effic. with excellent design and average mgmt.	65	70	70	65	60	65

Source: Adapted from Charles M. Burt, "Trends Toward Efficient Low Energy Irrigation." Paper presented at "Aqua 83," Acapulco, Mexico, 1983 (San Luis Obispo, CA: Department of Agricultural Engineering, California Polytechnic State University, 1983).

The Role of Flexible Water Deliveries in Water and Power Efficiency

Irrigation district canal and pipeline systems are typically designed for constant, steady-state peak flow rates. Delivery systems are rarely installed with sufficient attention given to the operational criteria for the unsteady or varying flow conditions that actually exist most of the time. With standard control procedures, farmers must submit a request for water delivery to the irrigation district at least one day in advance. The water delivery agency then computes the necessary discharge from the source and releases the water several hours or days in advance so that the required water reaches the user's turnout approximately when desired.

This method of operation (called manual upstream control) is inflexible. The abrupt change in flow rate made at the head of the canal must be continually adjusted manually to maintain a constant water level at the turnouts. This function can be automated, but even with automation, once water has been released into a canal, it must be used or else the far downstream end of the system will have spillage. On the other hand, if a farmer takes more water than scheduled out of a turnout, the source does not automatically and quickly increase the flow rate into and throughout the system; the downstream users end up with a shortage of water. Because of difficulties in maintaining constant water levels and balancing flow rates, farmers are not permitted to make unannounced changes in their flow rates or durations. In effect, they cannot "fine tune" their water deliveries and apply water efficiently.

Even with this rigid "supplier-oriented" operation procedure, flow rates through farmer turnouts vary constantly, making precise on-farm irrigation management very difficult if not impossible. At the far ends of many irrigation canals, the water supply is often available on a "feast or famine" basis.

These standard water delivery methods and practices are incompatible with advances in modern on-farm irrigation technology, which are capable of high irrigation efficiencies.[9] A report dealing with agricultural water conservation in California came to the following conclusion. "On-farm water savings can be best achieved by proper management of existing and new irrigation systems and through good irrigation scheduling programs which determine the correct timing and quantity of water application."[10]

----

9. See California Department of Water Resources, Water Conservation in California DWR Bulletin 198-84, 1984.

10. D. C. Davenport and R. M. Hagan, "Summary and Conclusion: Agricultural Water Conservation in California, with Emphasis on the San Joaquin Valley." Department of Land, Air, and Water Resources, University of California, Davis, California, 1982.

Within some irrigation districts, inflexible water deliveries by the district may be the single greatest hindrance to achieving high on-farm irrigation efficiency.[11] New irrigation methods such as drip, linear moves, and controlled volume furrow irrigation are very promising, but all require automation or the ability to control water supplies. A study of energy use for irrigation in the United States concluded that inefficient pumping and energy load management at the farm and irrigation district levels were significant.[12] Where water is supplied by districts, a flexible delivery system is necessary to be able to conduct energy load management at both levels.

Studies have shown that inflexibility of water delivery systems is due to both nonrecognition of the need for flexibility and to the lack of technology for implementation.[13] In a study of technologies affecting surface water storage and delivery, the U. S. Congressional Office of Technology Assessment stated that:

> Flexible delivery schedules are relatively new in concept, design, and implementation. For example, level-top and newer rapid-response downstream control methods remain experimental, design refinements are still required, and as yet, capital and labor costs remain high. The main advantage of these delivery schedules is the choice they provide in duration, frequency, and quantity of water delivered to ensure that the crop receives water when needed but not in excess of the amount required.[14]

Progress on Improved Water Delivery Methods

The Agricultural Engineering Department at California Polytechnic State University (San Luis Obispo), with the help of a $157,000 grant from Pacific Gas and Electric, has begun work on the construction of a major training, demonstration, and research facility for improved canal and pipeline automation. Expected to be completed within four years if the remaining funding is received, this facility will demonstrate a wide range of existing canal automation techniques. Using a 700-foot scale model canal with complete instrumentation, engineers will have the capability of verifying computer simulations of new and improved control methods. Research will continue on a promising canal automation technique developed at the university.[15] A pump station with a large

----

11. See C. M. Burt and J. M. Lord, "Demand Theory and Application in Irrigation District Operation," in the Proceedings of the ASAE Specialty Conference on Irrigation Scheduling, Chicago, IL, 1981).

12. See J. R. Gilley, et al., "Irrigation Management: Energy," in the Proceedings of the ASAE Second National Irrigation Symposium, Lincoln, Neb., 1980, pp. 127-140.

13. See C. M. Burt, R. W. Walker, and S. W. Styles, "Irrigation System Evaluation, 1985" (Department of Agricultural Engineering, California Polytechnic State University, San Luis Obispo, California, 1985).

14. Office of Technology Assessment (OTA), Water-Related Technologies for Sustainable Agriculture in U. S. Arid/Semiarid Lands. no. OTA-F-212 (Washington, D. C.: U. S. Government Printing Office, 1983).

15. C. M. Burt, "Canal Automation for Rapid Demand Deliveries (CARDD)" in the Proceedings of the ASCE Irrigation and Drainage Specialty Conference "Water Today and Tomorrow," Flagstaff, AZ, 1984, pp. 502-509.

combination of pump types and capacities will be used to demonstrate methods of automating pump stations to match flexible canal automation.

Hand-Move and Side-Roll Sprinkler Systems

Hand-move and side-roll sprinkler systems serve approximately 1,550,000 acres in California (15 percent of the irrigated acreage). Because most of these systems require pumps, energy conservation is important. There are two primary components of uniformity to understand: "catch can" uniformity, and flow rate uniformity.

Catch Can Uniformity. Figure 5.7 illustrates a theoretical grid between four sprinklers. The catch can uniformity (CCU) is a measure of how evenly water is distributed among the squares. Typical values of catch can uniformity range from 70 to 85 percent in new systems that have no plugging or sand wear. Factors that affect these values include wind, sprinkler pressure, sprinkler spacing, crop interference, nozzle size, sand wear, and plugging.

It is important to operate sprinklers at the proper pressure. A low pressure creates a "doughnut" pattern. The jet of water does not break up, and it therefore lands in a narrow ring, or doughnut, 30 to 40 feet away from the sprinkler. At high pressures, there is too much breakup, and there is excessive water application close to the sprinkler and susceptibility to wind distortion.

A management practice for improving catch can uniformity is the use of "alternate sets." By placing laterals in an intermediate location on every other irrigation, the spacing is effectively cut in half. The new catch can uniformity can be estimated as follows:

$$CCU_{\text{alternate set}} = 10 \sqrt{CCU_{\text{CCU original}}}$$

For example, if the original catch can uniformity was 75 percent, the use of alternate sets will result in an overall catch can uniformity (for 2 sets) of 87 percent. This technique is especially useful in windy areas. Sprinkler risers must be sufficiently high to put the sprinklers above the crop. If the sprinklers are too low, most of the water is distributed near the sprinkler.

Flow Rate (GPM) Uniformity. Catch can uniformity receives the greatest attention in most sprinkler evaluations, but in some cases the nonuniformity caused by unequal

Figure 5.7 Conceptual Grid of Catch Cans

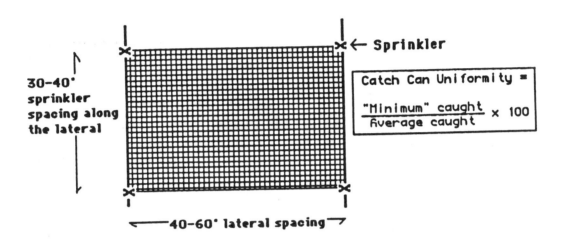

Note: The area depicted is between four sprinklers on a hand-move or side-roll system.

sprinkler flow rates is an even greater problem. The flow rate distribution uniformity (GPM DU) is defined as follows:

$$\text{GPM DU} = \frac{\text{"Minimum" sprinkler GPM} * 100}{\text{Average sprinkler GPM}}$$

The following factors affect flow rate distribution uniformity: pressure differences (elevation and friction), sand wear on nozzles, plugging, and different nozzle sizes used in a field.

Sprinklers will have a flow rate variation of 10 percent if there is a 20 percent pressure difference. In most hand-move and side-roll systems, there is considerably more than a 20 percent difference between sprinkler pressures. Pressure differences can be minimized by running laterals downhill when possible, using large pipe diameters, keeping reasonable lateral lengths, using pressure regulators on individual sprinklers, and using flow control nozzles on individual sprinklers.

Sand wear of nozzles and sprinkler drivers is a major problem in some areas of California, such as the central coast. The sand usually originates in wells. Good design of the well can minimize sand pumping. If a sandy well needs to be used, sand separators can be purchased to remove the sand. They can be installed either down in the well, at the suction of the pump, or at ground surface. These separators remove more than 90 percent of the sand and minimize sand damage. The submersible separators protect both the pump and the irrigation system.

An important factor when considering both catch can uniformity and flow rate uniformity is the fact that an irrigation pump supplies water to a complete field, not just to a single lateral of sprinklers. Therefore, the measurements must be taken throughout the field. Evaluations of uniformity in which only one lateral is analyzed will give results that are misleadingly high.

Pressure Requirements. Over the last five years there has been a drive to reduce the pressure requirements of irrigation systems. The energy cost is directly proportional to the pressure requirement, assuming all other factors stay constant (which is not always true). Low-pressure sprinkler nozzles have gained popularity in some areas. The objectives of switching from standard to low-pressure nozzles are to reduce sprinkler pressure from about 50 to 35 psi and to achieve good jet breakup and good catch can uniformity even at these lower pressures.

The reduction of sprinkler pressure by 30 percent will decrease the actual total pump pressure requirement by different amounts, depending upon the total pump lift.

Water Source	Total Pump Pressure reduction due to 30 percent sprinkler pressure reduction (percent)
Canal	23
Well lift of 100 feet	14
Well lift of 300 feet	8
Well lift of 500 feet	5

It can be seen that when pumping from a canal, the percentage of power savings is appreciable, but it is significantly less with a water supply from deep wells.

There can be potential disadvantages to using low-pressure nozzles. Some growers are very satisfied with the nozzles, but others have abandoned them because of several problems. First, large droplets are associated with lower pressures, which can cause sealing of the soil surface and increased runoff. Second, there are problems of lower flow rate uniformity, which will always be a negative result. The increased hours of operation necessary to offset this lower uniformity may negate any savings in horsepower. Third, pump efficiency may decrease. This result depends entirely upon the pump and the specific system. One application in which low pressure nozzle can virtually always be beneficial is in a system that is presently operating at suboptimal pressures.

Micro-Irrigation

Micro-irrigation is composed of drip (also called trickle) and micro-spray systems. These systems are characterized by very small flow rates at the emission devices themselves. The emitters are typically attached to a polyethylene hose of 1/2 to 1 inch diameter. Micro-irrigation was virtually unknown in California fifteen years ago; presently 300,000 acres, which is 3 percent of the irrigated acreage, are now irrigated with this method.

Many myths have arisen about micro-irrigation performance. As with any irrigation system, micro-irrigation has nonuniformities, losses, and potential for overirrigation. Because of the large amount of sophisticated hardware and moving parts with micro-irrigation, it is especially important to understand the systems in order to "tune them up" properly.

Uniformity. The following factors influence micro-irrigation uniformity (often called emission uniformity rather than distribution uniformity): clogging, manufacturing variation, aging of materials, pressure variations, unequal set times, and unequal drainage. A potential maximum uniformity of 90 percent was suggested earlier in this paper. Actual uniformities are, of course, considerably lower due to the factors just mentioned. The primary culprit for drastically lower uniformities is clogging of the small holes in the emitters. The hole sizes range from 0.02 to 0.06 inches in diameter, so one grain of sand can easily plug them. The problem is so severe that most large ranches with drip (micro) systems have personnel working full time during the summer driving along tree and vine rows to check for plugged emitters and to replace them completely.

There are several causes of emitter clogging. One is bacterial growth. Slimy bacteria grows inside the hoses and emitters. Large particles are sloughed off the walls, are carried into the emitter holes, and plug them. The source water can be dirty. Canal water can have everything from fish to clay. Well water usually has sand and often has oil and silt. Another cause is chemical precipitation. Carbonates of magnesium and calcium will precipitate and form a hard white deposit in the emitters similar to the hard deposit found in the bottom of old tea kettles.

Clogging problems can be prevented by a combination of good design and maintenance. Emitters should be selected which have large holes and short paths, if possible. Excellent filtration systems must be installed and operated properly to clean up the incoming water. A rigorous and systematic chemical injection program (using chlorine and/or acid) is a must. Hoses and submains need to be flushed on a regular basis.

Differences in flow rate between brand new emitters, all at the same pressure, can be a result of manufacturing variability. The smaller the hole in the emitter, generally the higher the variability. Drip tubing for row crops is more difficult to manufacture uniformly than orchard emitters, due to the very small size of the holes. Emitters with moving parts can also be difficult to manufacture uniformly. Moving parts also tend to have a second problem: their characteristics can change with time. A good manufacturer of micro-irrigation emitters will guarantee a minimum variability when the emitters are new and over a specified life of the system. Growers should insist upon such a written guarantee and should make the required measurements to confirm the guaranteed variation.

Pressure variations cause the same problems in micro-irrigation systems as in sprinkler systems, although some emitters are more tolerant of pressure variations than others. In an existing system, there are only two practical solutions to poor uniformity caused by pressure variations: install pressure regulators at submains or hose entrances, or adjust existing pressure regulators so that they all have the same setting.

On sloping topography, uneven drainage is a problem. At low points, some emitters run continuously, as they drain a complete mainline. This result contributes to poor uniformity and possible drowning of trees at the low points. There are three solutions to the problem. One is to irrigate with longer set times; the mainlines thus drain fewer times. Second, special non-drain valves can be installed. These valves are common in turf installations. They shut off the flow when the pressure drops to a preset value. Third, vertical pipe loops with air vents can be installed in the mainline and submains. The loops act as dams when the water is shut off and prevent line drainage.

In many orchards, especially with stone fruits (peaches, plums, nectarines, and so on), farmers plant 10- to 20-acre blocks of different varieties and continually replant as blocks get old. One drip system may supply many small blocks. There is usually more than one tree spacing among the various blocks. The emitters are often installed with the same number of emitters per tree, regardless of tree spacing. Each block is then irrigated the same number of hours. The result is a very different application rate per acre. For example, there would be a difference of 25 percent between a tree spacing of 16 x 16 feet as opposed to a spacing of 16 x 18 feet. The resulting nonuniformity can be solved by adjusting the hours of application to each block.

Drip irrigation has definite advantages in certain conditions. As with all methods, however, the system must be designed and managed properly to obtain reasonable irrigation efficiency. For example, it is just as easy to overirrigate with drip as with sprinklers. With either method, applying the water for an extra hour in a 12-hour day results in an 8 percent drop in efficiency.

Surface Irrigation

Surface irrigation includes such methods as furrows, border strip, and basin. In California, most surface systems use pumps; the pumps may be either on the farm or used by an irrigation district to supply the water to the farm. Surface irrigation is the single most important type of irrigation in California; 76 percent of the irrigation is done with surface methods. When properly designed and managed, surface irrigation can have very high uniformities and efficiencies; unfortunately, most surface systems are currently managed at suboptimal levels.

A basic difference between surface irrigation and micro or sprinkler irrigation is that with surface irrigation, the uniformity cannot be designed into the system. It is dependent on management, regardless of the location of pipes or canals and the size of those conveyance systems.

There are two other related factors to consider with surface irrigation. In most systems, at any moment during an irrigation, the volume of water which has and will run off the lower end of the field is unknown. The soil, rather than a pipe or air, is the transportation mechanism to individual plants.

With surface irrigation, the two factors, uniformity and average depth infiltrated, are dependent. This not the case with sprinklers or micro-irrigation; with these methods the uniformity (except for unequal drainage) will be the same for a 1-inch application as that for a 3-inch application.

In order to understand the justification for many improved surface-irrigation practices, an irrigation manager must understand more complicated principles than those for sprinklers. The hardware with surface irrigation is simpler, however.

Importance of Infiltration Rates. With sprinklers, systems are usually designed to apply water at less than the minimum intake rate that would occur during the irrigation event. As long as the intake rate is not exceeded, managers of sprinkler irrigation can assume that water infiltrates at a constant rate--the application rate.

Figure 5.8 illustrates a fundamental concept of water infiltration into soils. The depth infiltrated as opposed to the time of irrigation is shown for three hypothetical soils. There are several observations to be made. A 2- or 6-hour irrigation for one soil will not result in the same infiltrated depth as for the other soils. For most soils, the water infiltrates rapidly at first. As time increases, less water infiltrates per unit of time (that is, the intake rate decreases with time). The conclusion must be that with surface irrigation no two soils respond in the same way.

Figure 5.9 shows another interesting phenomena, which perplexes surface irrigators. Not only does the intake rate change *during* an irrigation, but also *between* irrigations.

Figure 5.10 illustrates the concept that by doubling the application time, twice as much water does *not* infiltrate. In other words, overirrigating by 10 percent extra time with surface irrigation does not necessarily mean an extra seep percolation loss of 10 percent. Instead, the applied water may run off and can be collected and reused. With volumetric control, this result does not apply.

Figure 5.11 illustrates the importance of having similar opportunity times for infiltration at various locations. The figure shows a "typical" California furrow irrigation situation in which water is run slowly down the furrows and then is shut off when the water reaches the end of the furrow. Water is shut off to prevent runoff (which is not wasted if it is collected and reused). The result is very different infiltrated depths at the ends of the furrows. In the figure, water sits at the top of a furrow for 7 hours, infiltrating 12 centimeters, providing 1 hour of opportunity time, and infiltrating 5 centimeters at that point.

With surface irrigation, the selected soil moisture deficit (SMD) at the time of irrigation will affect the distribution uniformity for furrows. Considerable water will infiltrate at the top end of a furrow before the water even advances to the lower end, resulting in poor distribution uniformity if the purpose is to irrigate an area of small soil moisture deficit at the lower end. If the soil moisture deficit is higher, the water must be applied longer, and the uniformity improves. Figure 5.12 illustrates irrigation of the same furrow at two different levels of soil moisture deficit. Both furrows have the same

Figure 5.8 Hypothetical Relation between Depth Infiltrated Versus Time for Three Soils

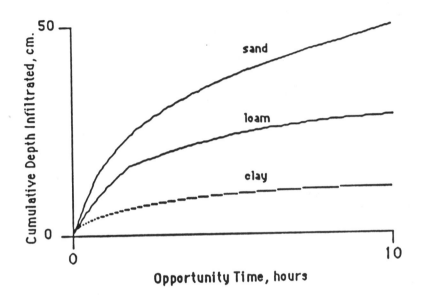

Note: It is assumed that the same percentage of the soil surface is flooded in all cases.

Figure 5.9 Infiltration for the Same Soil Throughout the Growing Season

Figure 5.10 Effect of Doubling the Opportunity Time on the Depth Infiltrated for a
Medium Textured Soil

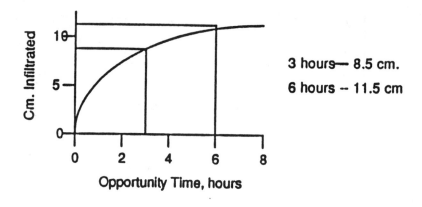

3 hours— 8.5 cm.

6 hours -- 11.5 cm

Figure 5.11 Effect of a Poor Advance Ratio on Distribution Uniformity

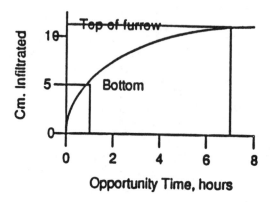

Figure 5.12 Irrigation of the Same Furrow at Two Different Soil Moisture Deficits

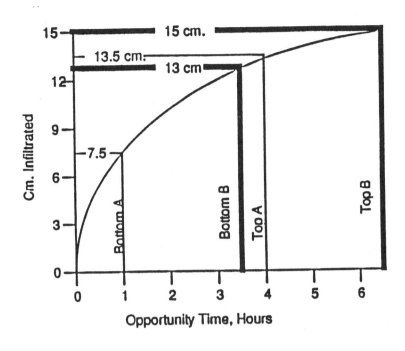

Figure 5.13 A Possible Field Configuration for Volumetric Control

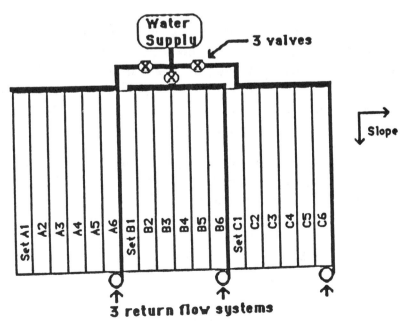

Note: The sequence of irrigating various sets is A1, B1, C1, A2, B2, C2, A3 . . .

advance time. Furrow A in the figure, with a small soil moisture deficit, has almost a difference of 100 percent in depth infiltrated at the top and bottom ends of the furrow. Furrow B, with a larger soil moisture deficit, has approximately the same depth infiltrated at the top of the furrow--13.5 centimeters in this example--as at the bottom end, which is infiltrated to 15 centimeters.

Although it is more difficult to measure uniformity with surface irrigation than with other methods, it is not more difficult to achieve good uniformity in a properly designed and managed system. The following factors are important: the water should have approximately the same opportunity time to infiltrate at all points in the field; runoff is necessary on sloping furrows; good land grading is essential; the greater the soil moisture deficit at the time of irrigation, the more uniform the infiltration can be (except for some cracking soils); and the furrows and border strips should be of a reasonable length.

How to Infiltrate the Correct Depth. As with other methods, achieving a good uniformity is only half the battle for high efficiency; one must also be able to apply the correct depth. It has been pointed out previously that it is virtually impossible for most irrigators to understand and obtain the equations that adequately describe the depth-time relationship. Therefore, surface irrigation requires some unique management. Specifically, a plan must be developed and followed for obtaining both high uniformities and the correct depth in the soil. Four strategies are listed below.

Depth of Infiltration Strategy #1: During the irrigation, probe the soil to determine how deep the water has gone. Before the water has reached the bottom of the root zone, shut the water off. Assuming one has followed the recommended practices of using a fast advance, a reasonable soil moisture deficit, and a return flow system, this technique gives both a high uniformity and the correct depth. The problems are that the irrigator must actually do some checking, and one does not know in advance what the irrigation duration will be.

Depth of Infiltration Strategy #2: Use a small computer program during the advance phase (while water is advancing down the furrow or strip) of the first set to predict the effect of different flow rates and set durations. Runoff is still required. Waterman Industries of Exeter distributes such a program at no cost. It is set up for the Apple II and IBM PC computers. Problems include the availability of a microcomputer, accuracy of data collection, and differences between wheel and non-wheel rows.

Depth of Infiltration Strategy #3: Use level furrows or strips. This technique requires laser land-grading and shorter furrows than those of sloping fields. But there is no runoff with level fields, so no pumps or return lines are required. The management involves applying the correct depth of water at reasonable rates. All of the water applied will infiltrate. The primary advantages of this strategy are that it is ideal for computerized irrigation scheduling and automation, and it is simple to understand. The primary disadvantage is that there is very limited range of soil moisture deficits that can be used to obtain a high uniformity. If the soil is not dry enough, the water cannot advance to the end of the furrows before it is shut off; if the soil is too dry, an excessive amount of water must be applied and the furrows may be inundated.

Depth of Infiltration Strategy #4: Use volumetric control on sloping furrows. The required system design is illustrated in Figure 5.13. This strategy is the best of all, because it is the simplest and, after some addition of sumps and return lines, it is adaptable to most existing fields. With this strategy, sloping furrows essentially become level furrows in their management. The technique is to apply a certain *volume* of water to a set of furrows or a strip. When that volume of water is applied, the water supply is switched to another submain. The sump in the field of the first set will collect and store the runoff. It can be switched on to return all the runoff to the *same area to which it was originally applied*. Therefore, the depth applied to a given area equals the depth infiltrated. By diverting the flow between three fields or sections, each sump has ample time to empty before another set in that block is irrigated. This strategy is ideal for automation of valves and fits into irrigation scheduling programs. Strategy #2 can be used to fine tune this method.

Major Causes of Low Surface Irrigation Efficiencies. Many commonly accepted "standard" surface irrigation practices can result in very low efficiencies. These practices include:

Not having a strategy for knowing in advance how much water will infiltrate during an irrigation. In other words, there is no good strategy for determining when to shut the water off. Presently inadequate strategies include the following:

Continuing an irrigation until water has "subbed" across the bed, as indicated by a completely dark soil surface. This practice is common even on sandy soils and results in tremendous water and fertilizer leaching.

Shutting the water off when it reaches the end of the furrow. This causes a large depth to infiltrate at the head. There is less infiltration in the middle, and the bottom end may be dry or flooded out, depending upon the slope and exactly when the water is shut off.

Operating on 12- or 24-hour schedules. A different depth infiltrates in a 12-hour set during the second irrigation than the third. Unless a rapid evaluation such as the Waterman Furrow Evaluation Program is run, the actual depth infiltrated is not known until after the field is irrigated.

Excess pre-irrigation. It is easy to overirrigate during the pre-irrigation because intake rates are highest at this time. This is the time, if any, to apply surge flow. Special care should be taken to run large flow rates with reasonable advances with all methods.

Excess first irrigations. The first irrigation (after planting) for many crops may only require 1 to 2 inches. However, a system may be incapable of infiltrating less than 3 to 4 inches at that time of the year. A good practice is to plan ahead and not fill the root zone during pre-irrigation so that the extra infiltration of the first irrigation will be stored in the root zone.

Lack of uniformity of flow rates and durations of sets. As an example, a field may really have four or five different set durations and flow rates during one irrigation. This can be caused by:

> **Accommodation of labor schedules.** Daytime irrigations may last nine hours, whereas nighttime irrigations are 14 hours long. Automation can solve this problem.

> **Improper use of tailwater systems.** In some systems irrigators will open more furrows or border strips when the tailwater begins to recycle back.

> **Improper use of cutbacks.** If flow rates are reduced when the water reaches the end, the cutback water must often be diverted to other furrows to prevent spillage from pipes or ditches. This is a system design and/or a management concept problem.

Loss of rainfall. In areas of rain during the growing season, it is often advisable always to underirrigate, leaving room in the soil to store unanticipated rainfall. With furrows, alternate furrows may be irrigated to accomplish this result in part.

Poor irrigation scheduling. Knowing *when* to irrigate together with knowing *how much* to apply requires knowledge of crop water use and soil water-holding capacities. This knowledge must be coupled with proper system design and management to infiltrate the correct depth.

Poor land-grading. Land-grading does not mean land-leveling. Slopes do not have to be constant or zero to be proper. Low spots that pond water, however, must be removed.

Trying to do the impossible. Surface irrigation performance on reasonable slopes and uniform soils can be superb. Steep slopes and non-uniform soils often warrant other methods of irrigation. Surface irrigation performance is also limited by the minimum depth of water that can be infiltrated evenly. If a crop is shallow rooted and needs a low MAD and many 0.5- to 1.0-inch irrigations, another irrigation method may be more suitable.

Surge Flow. Surge flow is a new concept in surface irrigation that has been prompted by both industry and researchers. It is defined as "a series of non-continuous pulses of water applied to a furrow or border strip." Typically, the water is applied in 4 to 5 pulses of 30 minutes to 2 hours each in order to advance the water to the end of a field. Upon arriving at the end of the field, the pulse duration is shortened. Equipment is readily available from all major surface irrigation equipment manufacturers to perform the automatic switching of flow.

On many soils, the surging action produces an accelerating sealing of the soil surface during water advance. This result can be beneficial in California for pre-irrigations, when soils are very loose and the intake rate is very high. In later irrigations in California, however, the soils are often sealed up excessively with regular irrigation, so this may do more harm than good. In any case, the automatic cycling after the advance will reduce runoff by effectively cutting the onflow in half. This result is very beneficial, as it reduces the required sump and pump capacities.

Surface irrigation has unique management requirements. There are four basic rules. Irrigate different soils with different durations. Have good land-grading. Irrigate so that opportunity times for infiltration are similar across a field. Develop a strategy for knowing the depth infiltrated.

*An earlier version of this paper appeared as "Water and Power Efficiency in Irrigation" in Daell D. Zimbelman, ed., *Planning, Operation, Rehabilitation, and Automation of Irrigation Water Delivery Systems*, proceedings of a symposium sponsored by the Irrigation and Drainage Division of the American Society of Civil Engineers (ASCE) and the Oregon Section of the ASCE, Portland, Oregon, July 28-30, 1987 (American Society of Civil Engineers, 1987).

Seasonal Summary Table

Crop	Area Acres	Expotransportation AF/Acre	Expotransportation AF	Effective Precipitation AF/Acre	Effective Precipitation AF	Crop Water Requirement AF/Acre	Crop Water Requirement AF	Irrigation Water Applied AF/Acre	Irrigation Water Applied AF	Irrigation Water Beneficially Used AF/Acre	Irrigation Water Beneficially Used AF	Seasonal Application Efficiency Percent	Seasonal Irrigation Efficiency Percent	Seasonal Underirrigation AF/Acre	Seasonal Underirrigation AF	Seasonal Deep Percolation AF/Acre	Seasonal Deep Percolation AF
Cotton	11235.0	2.0	22688.1	0.2	2454.5	1.8	20233.8	2.6	27655.7	1.6	20202.8	77	65	0.3	2950.0	0.8	9236.4
Tomatoes	1783.8	1.6	2854.9	0.2	329.1	1.4	2525.7	2.1	3810.4	1.6	8331.7	70	66	0.2	278.1	0.7	1177.5
Alfalfa Seed	1314.0	3.3	4386.2	0.3	368.2	3.1	4018.0	1.9	2492.1	1.5	2011.6	168	81	1.5	2006.4	0.3	450.8
Melons	261.2	0.9	227.2	0.2	51.7	0.7	175.5	1.3	336.0	0.9	785.1	54	63	0.0	0.0	0.5	118.5
Sugarbeets	707.0	2.7	1876.8	0.2	164.3	2.4	1712.5	3.0	2118.5	2.0	1911.3	84	69	0.4	297.1	0.9	613.0
Onions	40.0	1.3	53.3	0.4	15.3	1.0	38.0	2.8	111.3	0.9	36.7	34	33	0.0	1.3	1.8	71.3
Almonds	373.0	2.9	1080.0	0.2	65.5	2.7	1014.5	2.2	804.3	1.7	549.6	160	85	1.0	364.9	0.4	148.2
Bell Peppers	194.0	2.3	436.5	0.3	50.5	2.0	386.0	2.6	495.6	1.8	319.6	80	63	0.3	66.4	0.8	152.5
Barley	80.0	1.1	90.7	0.3	24.0	0.8	66.7	1.7	134.7	0.8	60.0	50	44	0.1	6.7	0.9	72.7
Weighted Means		2.1		0.2		1.9		2.4		1.6		86	67	0.4		0.8	
Totals	15988.0		33693.6		3523.1		30170.5		37958.6		34308.2				5971.0		12041.0

Note: 57 fields had beneficial use values assigned by the program advisor which were more than 15 percent greater than the Westlands Water District estimate of water requirements.

Those fields were not included in this table. This summary include no fields of garlic, lettuce, or cauliflower.

Equations used:

Seasonal Application Efficiency = $\dfrac{\text{Evapotranspiration} - \text{Effective Rainfall}}{\text{Irrigation Water Applied}} \times 100$

Seasonal Irrigation Efficiency = $\dfrac{\text{Irrigation Water Used by Crop}}{\text{Irrigation Water Applied}} \times 100 = \dfrac{\text{Beneficial Use}}{\text{Irrigation Applied}} \times 100$

1987 Westside Watershed District & Westside Resource Conservation District
Combined Data - Summary Table

Category	Sub-Category	Number of Fields	Acres	Irrigation Efficiency (Percent)	Distribution Uniformity (Percent)	Deep Percolation (AF/AC)	Deep Percolation (AF)
Annual							
Cotton		100	13,000	67	71	0.8	10,442
Tomato		40	5,335	74	75	0.5	2,868
Other crops		50	5,489	68	72	0.7	3,972
All acres		190	23,824	69	72	0.7	17,283
Pre-irrigation							
Short furrow		82	9,477	64	67	0.3	3,009
Long furrow		48	6,862	58	64	0.3	2,320
Sprinkler		53	7,148	77	64	0.1	661
Cotton		100	13,000	63	65	0.3	3,987
Tomato		40	5,335	77	62	0.15	791
Furrow with Tubewell		67	8,118	69	66	0.8	2,363
Furrow without Tubewell		63	8,221	54	65	0.4	2,966
Regular Irrigation							
Short furrow		134	16,654	72	76	0.5	7,758
Long furrow		40	5,624	70	74	0.5	2,582
Sprinklers		9	1,210	66	68	0.6	746
Cotton		100	13,000	69	74	0.8	10,441
Tomato		40	5,335			0.5	2,868
All other crop		50	5,489			0.7	3,974
Furrow with Tubewell		98	12,669	75	79	0.4	5,572
Furrow without Tubewell		76	9,609	68	71	0.5	
All acres		190	23,824	72	74	0.5	11,086

Regular Irrigation Irrigation Efficiency - All crops.

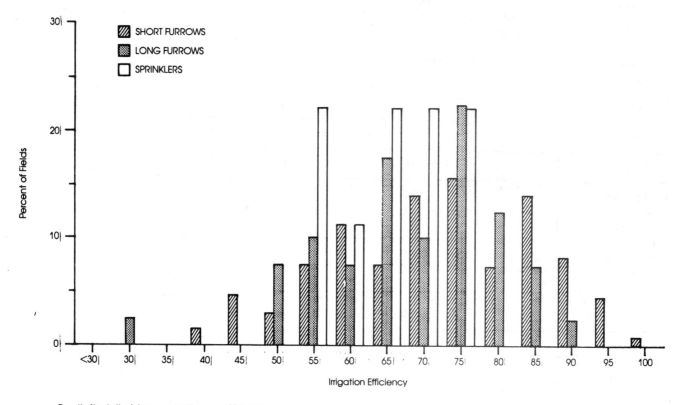

Results (tentative) from evaluations on 190 fields in
Central California Summer 1987.

Sample CARDD computer simulation results. Both runs show
good control of upstream and downstream water levels, plus
effective dampening. Note that the flow rate changes are large.

POOL DEPTH VS. TIME GRAPH

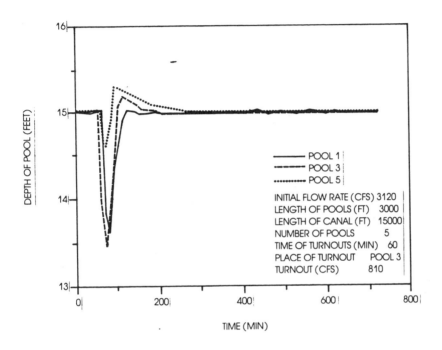

POOL DEPTH VS. TIME GRAPH

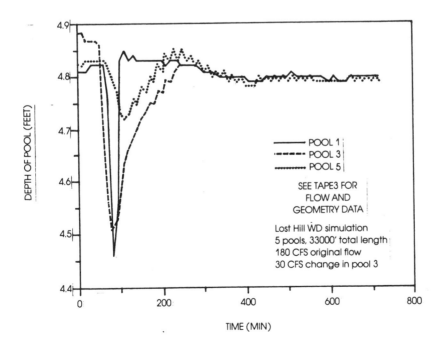

Chapter 6

DEVELOPMENTS IN WATER CONVEYANCE AND WATER CONTROL IN FRANCE

G. Manuellan

Immediately following World War II, it was thought that 2.5 million hectares were irrigated in France, which would have represented over 7 percent of the 34.7 million hectares of total agricultural area at that time. But these statistics were based on very unreliable estimates, and their economic significance was limited. For example, areas to which a water right was attached were counted even if the right was not exercised or if even only a small part of the irrigable area that would benefit was actually irrigated. The result was a substantial waste of water and low-density irrigated farming practices in most of the irrigated areas. In the old irrigation schemes, the volume of water drawn often exceeded estimated optimal amounts by two or three times. In addition, more than 1 million hectares of permanent natural grazing land, most of it in mountain areas, was irrigated fairly regularly. These irrigation systems were rapidly disappearing, however, because of the exodus of the population from these regions, the intensification of fodder production, changes in livestock systems, and so on.

The irrigation systems were old; some dated back several centuries. Farmers and public authorities preferred to give priority to investments considered to offer more immediate return, such as mechanization of agriculture, land consolidation, fertilizer use, processing of agricultural products, and potable water supplies for rural areas. Paradoxically, in the relatively dry Mediterranean region, the prosperity of vine cultivation, for which irrigation is not essential, did not favor development of irrigation. Irrigated areas were devoted to vegetables, fruit and flowers, alfalfa, and other products such as maize and tobacco that supplied the limited needs of the domestic market. A singularly important development of collective irrigation areas at that time was creation of the rice lands of the Rhône delta (Camargue) to make up for the deficit in imported rice during the war. Areas such as the Costière du Gard in the Languedoc region that are rich today were almost uncultivated in the 1950s, partly because of the lack of irrigation but also because of the limited economic prospects offered by outdated local land-ownership structures and the lack of an "agricultural project" mentality at that time.

New Technical Designs in Irrigation Projects

Beginning in the 1955 to 1960 period, irrigation in France benefited from new developments and underwent exceptional technical renewal. Irrigated areas were expanded, and several old systems were modernized. The old irrigated systems had remained virtually unchanged since they were first built. Traditional surface canal or water-tower irrigation gave way to the general model of irrigation under pressure, on demand, with sprinkler application and metering of water volume. In the initial stage, the new systems had rotary sprinklers mounted on light movable or semi-fixed piping, with a spacing of 12 to 24 meters; the systems had low or medium pressure (about 3 bars). Application rates were low, about 5 to 7 millimeters per hour, which adapted well to a wide range of soil permeability and crops.

About 1970, this single model was diversified with the use of sprinkler systems with varying characteristics, particularly irrigating machines that were usually energy intensive but labor saving, and with the use of localized irrigation or micro-irrigation methods. Since 1980, regulated and automated control of water flow--already a feature of the new irrigation networks in the form of free-flow canals, pressure piping, pumping stations, and so on--was extended to include automated command of water application at the field level, a development associated with labor shortages and high wages and with improved farmer skills.

Recently, however, there has been a resurgence of interest in traditional surface irrigation systems. Rather than treating these systems as doomed to extinction, merely waiting for the transition to a modern total sprinkler irrigation scheme, designers have sought less radical ways to improve them that would require less capital and energy. New procedures and materials are being studied that would, through automation, reduce constraints and high labor costs and improve the performance of the old surface irrigation systems that remain in operation in the Mediterranean region of Provence.

Throughout all the recent technical developments, the choices made were not based on "best available technology" criteria. In assessing the advantages and disadvantages of the available techniques, planners considered their adaptation to the environment and their capacity to eliminate obstacles as the main determining factors. To ensure a reasonable return on the new collective investments and justify their implementation, they sought first of all to find ways of rapidly increasing the area actually irrigated. In old schemes with traditional water-tower surface irrigation, the considerable time lag between plant installation and actual use of the water by farmers was economically unacceptable. In the new schemes, the design eliminated the rigidity of the water tower and the costs of land preparation and leveling, which farmers were always reluctant to incur.

Since wages were rising more rapidly than other factors of production, the old surface irrigation designs, which were very labor intensive and inefficient in water utilization, had to be renewed. The configuration and scale of the projects had to be diversified to allow expansion of "supplemental" irrigation into subhumid areas. The contingent nature of supplemental irrigation meant that it could not be handled by the water tower, however, and land could not be taken out of production for canal network rights-of-way and permanent structures. These obstacles could be avoided only by using sprinklers and movable piping (and nowadays irrigation machines). Given the problem of water rights and the competition from other economic activities for the water resources, designers optimized the economic utilization of the investment by introducing a combination of sale of water by volume, use of a water-saving irrigation method (sprinklers), and inclusion of irrigation in multipurpose projects whenever possible.

Land use that is flexible for agriculture and other activities meant that permanent structures on the land had to be limited, and construction of canals and surface works had to be minimized. Water had to be transported in buried pipes under pressure. More than in the past, water management had become a major concern. From the outset, the design had to be linked more closely to management to ensure that the works would be easy and economical to manage and that they were suited to the farmers' real needs. In addition, various institutional measures, both existing and newly created, had to be taken into consideration. Suitable supervisory units had to be capable of resolving problems of financing for major investments, tariff setting and cost recovery, ensuring operation and maintenance, providing technical assistance to farmers, and so on. The regional development companies were established in 1955.

The choice of irrigation technology was highly compatible with the concurrent development of industrial production capacity, the service capacity of commercial firms, and the construction capacity of French contractors. Analysis of all the factors in the technical developments that took place during this time suggests that the irrigation designs in France corresponded to what is today termed appropriate technology in terms of both inherent innovation and transfer of imported technology.

Modern Irrigation Systems

The total area irrigated in France at the end of 1987 was estimated to be 1.7 million hectares, of which 150,000 hectares have old surface irrigation systems, 1.5 million hectares have post-1955 sprinkler irrigation, and 50,000 hectares have micro or localized irrigation. The area equipped with modern irrigation since the mid-1970s has increased on average by 50,000 hectares a year. Micro-irrigation area alone has increased from 2,000 hectares in 1974 to 50,000 hectares in 1987.

Farm-Level Sprinkler Irrigation Systems

Traditional nonmechanized sprinkler irrigation systems are still fairly common in France. There are three types: partial coverage, full coverage, and integral coverage. Partial-coverage systems have movable piping equipped with either rotary sprinklers under moderate pressure (2 to 4 bars) or rainguns under higher pressure (up to 7 bars) spaced farther apart (36 x 36 meters). With full coverage, only the sprinklers are moved to each position. This category includes sprinkler systems mounted on a chassis (move) connected to the hydrant by flexible pipe. Each move allows irrigation of 5 to 7 positions from the point of connection to the hydrant. In integral coverage systems, the entire area is covered by standpipes with sprinklers or rainguns, which remain fixed throughout the season. On land with perennial fruit crops, the pipes can even be buried. The system has high capital costs but is labor saving. Command of irrigation can be automated.

Of the irrigating machines in use, giant irrigators are becoming increasingly rare because of their sensitivity to wind, the difficulty of moving them because of electricity lines, high pressure, and so on. These large machines have revolving arms equipped with several booms irrigating in fixed position and supported on a self-propelled or tractor-drawn chassis for travel from one station to the next.

Other types of irrigating machines that irrigate adjacent strips of land by means of a sprinkler unit that moves automatically and irrigates as it goes include **travelers** (machines advanced by cables) and hose-coilers, the type most commonly used in France. Hose-coilers are easy to use and are employed in major irrigated farming, including vegetables on small plots and maize or forage crops on large areas of land. On some of these machines the winding drum is fixed at each irrigation position, and the sprinkler gun is pulled along by its polyethylene feed pipe. On other machines, the winding drum and the gun are located on the same chassis, which pulls itself on the feed pipeline. There are also self-propelled hose-coilers, which can move independently from one work position to the next.

These machines require a relatively high entry pressure: 6 to 7 bars, or up to 10 bars for the very large units. The movement of the raingun and the winding of the pipe on the drum are driven hydraulically; depending on the model, they are powered by either a piston, a bellows jack, or a turbine run by the irrigation water under pressure at the entry. These machines are equipped with various safety and regulating devices that ensure, within reasonable limits, regular speed of advance of the raingun, homogeneous irrigation, and regular winding of the pipe on the drum.

Pivoting bars consist of towers mounted on wheels and powered by an electric motor. The spans between two neighboring towers, spaced 30 to 70 meters apart depending on the situation, consist of a metal-frame beam. The unit rotates around a central pivot that is connected to the water feed point. There can be 1 to 20 spans.

The water line, independent of the metal-frame beam, consists of jointed elements to the right of each tower so that they can form a specified angle between them. Sprinkling devices (sprinklers or spray booms) are arranged on the water line, suitably spaced so that the application rate is the same over the entire irrigated area. The end span often terminates in a bracket that carries an end raingun. Supplemental units (corner systems) can be added to irrigate areas outside the circle. Coverage independent of the pivoting bar is also used for these areas.

The pressures required at the central pivot vary with the length of the pivoting bar, the flows to be provided, and the sprinkler service pressures (2 to 4.5 bars) or spray boom pressures (1 to 1.5 bars). The end raingun requires a pressure of 3.5 to 7.5 bars. The pressures at the central pivot thus range in practice from 5 bars for a spray bar of 300 meters in diameter to 10 bars for an 800-meter bar.

The pivoting irrigation bars are equipped with various safety and automatic alignment control systems. Rotation is regulated on the basis of the forward progress of the end tower and for each intermediate tower that is ahead of the tower downstream.

These appliances are now very reliable. They adjust well to varied topographic and agronomic conditions and to most crops. Their capital and labor costs make them very competitive with hose-coilers, and the costs are only one-half those of conventional integral coverage. Capital costs per hectare decrease appreciably with the length of the pivoting bar.

Frontal bars with continuous travel move linearly and irrigate a rectangular area. Some are able to pivot on themselves at the end of the irrigated strip to provide water to the adjacent strip; this arrangement is a combination of the frontal bar and the pivoting bar. The quality of irrigation is further improved by increasing the number of low-pressure distributors on the bar and bringing them close to the ground by means of small vertical branches. A mechanized system of localized irrigation can be devised by suspending polyethylene pipe with calibrated nozzles on the under side that deliver small flows at actual ground level in the furrows.

In the automation of farm-level sprinkler systems, irrigation can be controlled with a tensiometer; with the sensors gauged properly, irrigation can begin at the proper time and its efficiency can be controlled. It is also possible to computerize automatic irrigation with "Quadrimatic" high-pressure sprinkler systems equipped with rainguns spaced 90 to 100 meters apart. The farmer selects the time and duration of irrigation for each raingun, and this information is stored in a central coding unit and a decoding "pilot" for each raingun, which are connected to each other by a 2-wire electric cable. The cable supplies electricity to the pilots and transmits the orders to open and close each raingun feed valve after decoding by its "pilot." The computer can control 128 pilots (or groups of pilots) and run them in sequence for an irrigation time of 0 to 99 hours and use 99 different cycles. A sensor, such as an anemometer or a rain gauge (which would shut off irrigation), can control the central unit. Two different irrigation programs and one fertilizer program can be recorded in the central unit.

Localized Irrigation or Micro-Irrigation

In localized irrigation, water is supplied in small but frequent doses at ground level without wetting the foliage; only part of the soil surface at the base of the plant receives water. Part of the volume of soil within the root zone, known as the "bulb," is kept at a humidity level that is close to the soil's retention capacity. The equipment required for the system includes a pressure regulator, a flow limiter, a water meter, a filtration unit, which consists of one or more sand or screen filters, and a fertilizer injector (or a liquid fertilizer dosing pump). Water is distributed via a network of polyethylene or PVC surface pipe, which operates under low pressure (class 1, 4 bars), and ends either in tubes or distribution bars with a diameter of 10 to 33 millimeters. These tubes are either porous piping or perforated simple or double-tube pipe.

The distributors in a localized system differ according to the way water is used. Point localization uses drip feeders, with a flow of 1 to 12 liters an hour. Line localization uses brass nozzles having a diameter of 1.2 to 2.1 millimeters, with a flow of 45 to 100 liters an hour, that are fixed on a bar placed in the bottom of a narrow furrow, which is blocked off into sections or bays--the *Bas Rhône* (Lower Rhône) procedure. In small-area localization, mini-distributors (fixed jets) or micro-sprinklers (rotating jets) are used; the flow ranges from 30 to 150 liters an hour. The first of these, the drip system, is the most versatile one. The other two are used primarily for fruit and vine crops. The Lower Rhône procedure can also be used on vegetable crops. Sugar cane is irrigated by the drip system as well as the double-tube method; the unitary flows of drip-feeders make them susceptible to clogging.

Localized irrigation has several advantages. Irrigation is precise; all the water is distributed to the plants. There is no wind sensitivity. The system saves labor, energy, and water. Fertilizer can be applied with the irrigation water, particularly in the drip method. The system is easier to use than others, and less cultivation is needed. Fields can be cultivated without interfering with the irrigation timetable. The soil structure is better maintained. The system can readily use low flows, water that is high in mineral salts, and, with good filtration, waste water. Steep slopes can be irrigated. But the system does have disadvantages, primarily in its high capital costs and its susceptibility to obstruction and clogging. Close attention has to be paid to ensuring continuously sound performance of the filters, descaling and flushing, and watching over the functioning of the drip-feeders and the entire installation, line by line. More than any other method, localized irrigation requires top-quality design, materials, and management.

The conditions of spatial diffusion of water into the soil from the localized water injection points and the formation of the moist "bulb" in the root zone pose problems that remain to be solved. In excessively permeable or stony soils, or soils too rich in fine or swelling elements, lateral migration of the water is difficult, and the humidity profiles do not cover the whole of the root profiles. To ensure proper supply to the plants, designers can increase the density of the drip-feeders, at least to the limit imposed by the incremental cost. In addition, irrigation can be started early in the season, which prevents drying out of the soil and facilitates rehumectation. In swelling soils with shrinkage cracks, drip-feeders can be replaced by mini-distributors or micro-sprinklers.

Solutions do exist, but the problems have to be worked out comprehensively; local experimentation is often necessary for determining the optimal design characteristics for individual plots--the arrangement of crop lines and bars, drip-feeder characteristics, flows and distribution, irrigation dosing and intervals, fertilizer applications, and so on.

Localized irrigation lends itself well to automated management. The farmer has many practical reasons that justify automating the network: to control irrigation timing and location, to manage large numbers of plots and irrigation positions, and allow parcel fragmentation. Irrigation can be automated in two ways--volume control and time control. Volume control is based on the use of volume valves combined with hydraulic valves. They allow delivery of a predetermined volume to each irrigation position and automatic switching of the flow from one position to the next, with as many successive positions as desired. Some volume valves can be reactivated automatically at the end of the irrigation cycle.

Time control allows irrigation to start and stop at predetermined times by means of a built-in clock. The electromechanical programmers, or preferably electronic programmers, can be either individual, integrated into the control valve, or central, in which case they control several positions. Special irrigation computers can add various automated functions, such as control of fertilizer applications, cleaning of filters, starting and stopping of pumps, and record keeping of irrigation episodes by storing in the memory the quantities of water used at each irrigation. Automatic command of irrigation episodes is controlled by climatic parameters via the soil (using such devices as tensiometers or electrical resistivity probes) and the plant itself. In the latter case, delicate sensors measure the difference between air and leaf temperatures or, alternatively, dimensional differences of the fruit or stalk between day and night when the plants lack water (hydric stress of the plant).

Although automatic irrigation at plot level is still in the incipient stage, it is already playing an important role in increasing agricultural productivity and the efficiency of water and fertilizer. This growing trend requires a comprehensive approach, from design to management, in solving the problems. What is needed is research and development of suitable materials, the acquisition of technical and economic reference data, project design, the training of farmers and staffs of irrigation management agencies, and attention to the management of irrigation systems, network operation and maintenance, on-farm conduct of irrigation, and technical assistance to farmers.

Modernization of Gravity Irrigation Systems

Traditional surface gravity irrigation systems can be modernized without converting to sprinkler or drip irrigation. Such modernization can reduce labor costs by reducing the number of hours and eliminating night and Sunday work, improve water economy and irrigation efficiency, and increase production. In some old irrigation systems still in use in Provence, the water quantities drawn at the offtakes often exceed the water requirements of the crops by five times, and water doses per irrigation episode are 150 to 200 percent of mean evapotranspiration. Systems can be modernized to keep investment costs within moderate limits and lower the cost of energy.

Even today, surface irrigation can be relatively attractive in certain areas if modernization can be done at low cost and the labor constraints can be overcome. Provence is one of such regions; a survey of the costs of various irrigation methods to farmers was conducted in 1981 (see Table 6.1). While researchers took care to make the calculations comparable, the results have to be interpreted with great caution, since it remains difficult to compare surface, sprinkler, and drip methods. In Provence, the surface networks are very old, their efficiency is low, and efforts to limit labor costs are detrimental to water-use efficiency. The discrepancies in collecting water tariffs in the different management agencies are considerable; the price of water in the old surface networks barely covers collective operating costs and minimal maintenance of the works.

In the 1980s, new materials were developed and adapted to specific plot conditions, particularly to improve water distribution at the head of the plot and to automate distribution, making it more precise. Theoretical research is being pursued to improve mathematical models of surface irrigation parameters (irrigation scheduling) and to make practical use of them. The new materials and techniques have been aimed particularly at easing the constraint of a low water head--generally below 0.5 meters of water--in gravity systems without pumps. At the head of the plot, valved pipes ensure satisfactory uniformity of the flows distributed in each furrow. They consist of inexpensive adjustable valves and light PVC piping of 150 to 200 millimeters in diameter, treated against ultraviolet radiation, with elastic joints, which are easy to disassemble. Flexible tubing is made of plasticized PVC. In the "Transirrigation" system (the "cablegation" system in the United States), a piston moves at a regulated speed with the perforated distribution piping placed at the head of the plot. The movement of the piston from upstream to downstream of the piping places successive groups of orifices under the water line for a predetermined duration. Each successive group of orifices thus supplies the corresponding number of furrows, which represents an irrigation position. Miscellaneous valves have been developed that ensure automatic or semiautomatic flow transfer from one irrigation position to the next.

An improvement and demonstration program that covers all these materials and techniques is being conducted to obtain more precise figures and to demonstrate the value of these innovations to farmers in areas where surface irrigation systems can be

Table 6.1. Survey of Comparative Irrigation Costs in Collective Systems

for Farmers of Provence (Mediterranean Region), 1981

Irrigation technique	Pasture, maize, fodder crops	Fruit trees	Vegetables, truck crops
	(francs per irrigated hectare per year)		
Surface irrigation	1,100-1,600	1,100-1,600	1,600-3,300
Sprinkler irrigation	1,600-2,500	1,750-2,550	1,750-2,450
Drip irrigation		1,800	3,600

Notes: Capital charges are not comparable in the collective networks; they are practically zero in the old surface irrigation systems.

Costs are calculated as follows:

For traditional surface irrigation systems:

Labor charges for irrigating at the farm level + labor charges for maintenance of on-farm canals and restoration of plots.

For sprinkler and drip irrigation systems:

Depreciation charges on irrigation equipment (relatively sophisticated equipment of the type in use in Provence) + purchase price of water from the various collective network management agencies set in the 1980s.

modernized. Other methods being introduced are land leveling with the use of rotary laser-guided earth-moving equipment, which has proved to be accurate and reasonable in cost, and intermittent-flow furrow irrigation, or surge flow irrigation. This latter technique has not produced conclusive water savings in the French setting, where furrows generally do not exceed 100 meters in length.

Costs of Irrigation Systems for Different Crops
Indicative Investment Levels on January 1, 1982

Equipment and Density	Crop types	Price range (Francs per ha including taxes)
LOCALIZED IRRIGATION		
Perforated bars, 400-800 nozzles/ha with filter and pressure reducer, for 2-3 ha	Fruit crops	7,500/ 9,500 (8)
Drip 800-1,000 drip-feeders/ha, with filter and pressure reducer, for 2-3 ha		7,000/ 9,000 (8)
Drip 12,000-15,000 drip-feeders/ha with filter and pressure reducer	Vegetables, flowers	20,000/30,000
SPRINKLERS		
SPRINKLERS:		
6 x 12 m per position covering 1/10th of area	Open-air	4,500/ 5,400
6 x 12 m per position covering 1/5th of area	vegetable	5,700/ 8,300
6 x 12 m with flexible pipe allowing 5 positions	crops	8,500/ 9,500
6 x 12 m total coverage		21,700/26,500
6 x 18 m with flexible pipe allowing 5 positions		5,800/ 7,300 (2)
OSCILLATING BARS:		
Partial coverage of 1/4 of area		9,000/12,000
Total coverage		28,000/35,000
SPRINKLERS:		
18 x 18 m per position covering 1/10th of area	Fruit crops	2,500/ 3,200
8 x 18 m per position covering 1/5th of area	Forage crops	4,500/ 5,000 (1)
	Maize	4,900/ 5,400
	With rod:	8,500/15,400 (5)
8 x 18 m total coverage, piping, 1/5th sprinkling		10,000/17,000 (7)
8 x 18 m total coverage		3,900/ 4,400
8 x 24 m per position covering 1/5th of area		4,000/ 4,500
	With rod:	9,000/15,000
4 x 24 m total coverage		
RAINGUNS: 38 x 38 m per position, covering 1/5th of area		7,000/ 9,000 (6)
COMPLETE GRID: Total coverage, 33 diameter pipe sprinkler movements		5,500/ 7,500 (4)
HOSE-COILERS: 63 or 75 diameter pipe, flow 12-30 m³/ha, for supply of 1,200 m³/month, area irrigated 5-20ha		7,000/4,000
82 or 90 diameter pipe, flow 30-80 m³/ha, for supply of 1,200 m³/month, area irrigated 10-25ha		7,000/3,500 (3)
GREENHOUSES		
SPRINKLERS mounted in 3 x 3 m or 4 x 4 m: Aluminum piping	Vegetable or	32,000/38,000
PVC piping	flower crops	29,000/35,000
(2)...(8) are the standard equipment types and investments identified in the following table, which also shows labor times for moving the equipment.		
PIVOTING BARS: 25-30 ha		6,000/ 7,000
150 ha		4,000/ 5,500

Note: The table presents very schematically, the various irrigation equipment possibilities and help the farmer determine capital cost and labor time for each irrigation set-up. Each type of equipment nevertheless remains a special case which must be studied in detail to ensure that it satisfies the farmer's purposes and constraints.

Transirrigation System of Furrow Automation
Used in France

System Principle

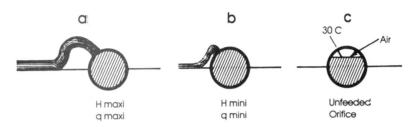

Longitudinal Section

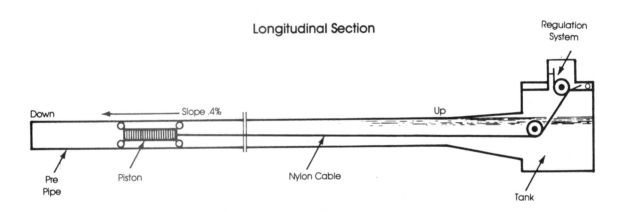

Plan View

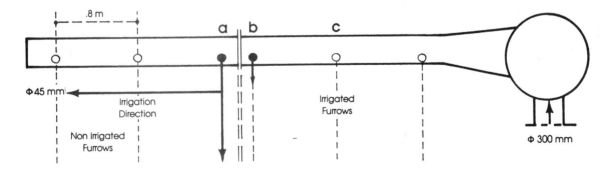

PART III

RESEARCH PRIORITIES

Introduction

The papers presented encompassed several aspects of research priorities: adaptating modern irrigation methods to research priorities of developing countries (D. Hillel), identifying research priorities and allocating resources (M. Jensen), and broadening the scope of the research effort (M. Ben-Meir).

Following the previous day's discussion, which concluded that more research is needed on irrigation and drainage and that the existing gap between irrigation and agronomy needs to be narrowed, Mr. Hillel started the session by describing the adaptation of a sophisticated, modern irrigation technique (drip) to replace more wasteful, traditional methods, using Baluchistan as an example. Under specific circumstances encountered there, some farmers are able to move from traditional flood irrigation to drip irrigation in four years. Drip is well suited to regions where water is at a premium. Whereas the emphasis previously was on soil moisture storage between discrete irrigations, it is now recognized that continuous optimal moisture in the root zone leads to the greatest plant yields. Advantages of drip irrigation are that it promotes favorable soil moisture conditions even in poorer soils; uniform application under uneven conditions of soil and topography; good aeration; free root growth; control of soil salinity, plant diseases, vegetative damage, and weed growth; and less evaporation and soil compaction. Problems remain with system management, such as the need for filtering to prevent clogging and salt accumulation on the periphery of the wetted root zone; the finely tuned management that is required leads to crops being far more vulnerable to disruptions in the irrigation regime.

Because of the much more sophisticated, precise management required, hasty adoption has led to disappointment in some cases. In particular, problems in adapting modern irrigation methods such as drip vary greatly with local conditions, and location-specific research therefore needs to be greatly stepped up. For proper control, good information is required. Monitoring is still inadequate, and workable procedures to measure potential evaporation are required. Since this information is unavailable in developing countries, the question is raised whether methods such as drip are too high tech for some of these countries. Should the present emphasis be on less-complex, more widely adaptable systems? For instance, bubbler systems, with wider orifices and lesser requirements for pressure and filtering, go part way toward this approach but at the expense of loss of control. The challenge is to develop a low-cost, low-volume, high-frequency system with as much control as possible. Such systems need to be sustainable; in the past, irrigation has too often been self-destructive.

Mr. Jensen in his presentation reviewed some basic principles, criteria, and procedures for identifying research priorities and allocating resources. To put this topic in context, he pointed out that the relationship between total crop yield and evapotranspiration, and therefore between yield and water-use efficiency, is linear. The most efficient water use is thus only possible when water is not limited. It has been possible to take advantage of improved irrigation methods as improved crop varieties have led to greater yields and efficiency of water use. This result has been due to a change in harvest index rather than in production of more dry matter. Genetic potential may well limit further increases, and future gains will not be easy.

To maintain gains made in high-level, efficient production, irrigation specialists need to avoid making mistakes such as those made in the past--for example, in neglecting requirements for drainage and more efficient conveyance and irrigation, which led to deteriorating conditions due to high water tables, waterlogging, and salinity. Conjunctive use and reuse of irrigation water sources need to be more widely considered.

With regard to research, it was pointed out that research *needs* are relatively easy to determine. Various activities carried out to define research needs, including the seminar conducted in the England on this topic in April, 1987, were described. Many needs have been identified, but they do not indicate priorities or the programs and resources required. It is much more difficult to set research *priorities*, which should encompass the mission and expertise of the research body, the resources required, and the need of the users for this research. Determination of research priorities requires that users as well as research leaders be involved. Other considerations are the required research infrastructure as well as the research approach. Also important is the level of research needed: whether it should address broader issues as part of a national research system, for instance, or whether more emphasis should be on adaptive and site- or project-specific research.

Previous efforts to identify high-priority research focused on the need for performance data and criteria and more adequate implementation and monitoring. In particular, site-specific and interdisciplinary research is needed. Agricultural research is generally better organized, with its established principles and procedures, than irrigation research, which has historically tended to be regarded as engineering. Research training and proper experimental design are inadequate, and statistical analyses are frequently lacking. People-related and intersectoral problems are often not appreciated. Better training and a broader intersectoral focus on interactions among plants, soils, climate, people, engineering, and management are needed. Each irrigation organization should have an adaptive research component receiving adequate, stable funding. It should also be remembered that irrigation management is part of an integrated production system.

In **Mr. Ben-Meir's** presentation, he described the process of establishing research priorities in Israel within the context of the development and adaptation of drip irrigation technology to the specific conditions in that country. In his opening remarks, he stated that the Bank should not be looking for universal answers in the question of technology, as these do not exist. He even expressed doubt as to the existence of universal questions. He observed that agriculture is dynamic, and one has to have a sense of destination. One must consider a very wide spectrum of factors when deciding on technology, which in turn has to evolve and become ever more sophisticated. An important area of the Bank's assistance should be in the establishment of an institutional system to deal with development of technology in countries where such systems do not exist. Many factors other than irrigation technology should be considered in setting research priorities. Particularly important are the environmental consequences and sustainability--not only within the perimeter of the project area but also beyond. The approach should be multidisciplinary, and targets in research should be an integral part of the general planning process. Requirements are location specific, and there is no global solution that can simply be copied from elsewhere. Each situation has to be specified according to its own characteristics, and both long- and short-term research has to be considered in total context in order to set priorities.

In Israel, priorities are conditioned by the following assumptions: water availability for irrigation will decrease; other uses of water will increase; pollution will increase; costs will rise; water quantity and quality will decline, and soils will be increasingly marginal in areas to be developed in the future. On the other hand, production and incomes need to increase to be able to finance increasingly more expensive water development. It is clear that this result cannot come from irrigated agriculture alone. In Israel, research efforts need to concentrate on the following: soil sanitation; environmental conditions to reduce damage and give maximum benefit; biotechnical research (for example, breeding for salinity resistance); increased water-use efficiency and reuse of water; and disease and pollution control. The most important considerations determining the Israeli approach are water scarcity, desert conditions, and saturated markets. Agricultural research is the backbone for the development of continuously more sophisticated technology and approaches that are able to pay the high cost of the infrastructure needed. It should be location specific and take a holistic approach.

Participants reached several major conclusions about research priorities. On the one hand, there is a lot of information and technology that is not being applied because technicians and farmers may be unaware of it. On the other hand, crucial knowledge in some areas is lacking, particularly information on performance as well as basic data needed for proper identification, planning, design, and implementation of successful irrigation schemes. There is a need for better communication and dissemination of available information as well as increased emphasis on research, particularly adaptive research at the system level. Opinion was expressed that procedures for agricultural research are better established than those for water.

It was noted that variation in performance is as great or greater within irrigation systems than between systems, which suggests that improved management of available technology is as important as the development and dissemination of new technology. The variation was not due to irrigation alone but to a host of other agricultural, managerial, and social factors. Irrigation schemes need to be planned and implemented as integrated production systems.

The research needed to maintain and develop technological improvement is carried out at different levels—international, national, and local—and support can be given to the general research systems or to more specific topics or projects. All are important, and the emphasis given in research support depends on different countries and situations. Indeed, the World Bank supports all levels, depending on particular needs. It is important, however, to give more emphasis to scheme-specific, adaptive, or operational research. The needs are very location specific; simple transfers of technology from one location to another is difficult.

Research needs are relatively easy to identify. It is far more difficult to set research priorities, and we should do a much better job here. We need to develop a better approach and process for setting priorities and to involve administrators, planners, and researchers, as well as the ultimate users. Priorities are very situation-specific, and therefore the institutional capability and competence to plan and implement what is necessary, particularly at the local level, is important and should be developed.

There was general consensus that one needs to look at irrigation schemes in a holistic way as an integrated production system. Research and technology need to be developed in a multidisciplinary manner, involving various specialists. So far we have not done a good job, although we are moving toward it. In particular, engineers, agronomists, and economists need to talk the same language. What is needed is a different professional preparation, much training and retraining, and attention to cross-sectoral issues.

Chapter 7

ADAPTATION OF MODERN IRRIGATION METHODS TO RESEARCH PRIORITIES
OF DEVELOPING COUNTRIES

Daniel Hillel

Traveling through the sparsely settled region of the Baluchistan Desert of Pakistan in 1982, I noticed an old farmer irrigating young apple trees. The trees were planted in basins, and the farmer was flooding the basins with an enormous excess of water. Such a waste of precious water seemed particularly unwarranted in so dry a region. Through an interpreter, I asked the farmer if all that water was necessary. He said, "Certainly. If we want the trees to grow big, we must give them much water." "Would you believe," I asked, "that big trees can be grown without flooding, by dripping water from narrow tubes, drop by drop?" The farmer chuckled in disbelief, winking at the interpreter and nodding his head at the ridiculous notion of the strange visitor. "No," he said, "that is quite impossible." "Well," I kept on asking, "would you believe that it is possible to take hundreds of people, put them in a container, lift the container into the sky, and then send it half way around the world?" "Of course," he answered without hesitation, "that's an airplane, that's science, and science can do many things!" "So why can't science irrigate big trees by giving the water one drop at a time," I kept pressing him. Again the kindly old farmer smiled indulgently, amused rather than annoyed. He pitied me for my ignorance. Once again he winked at our translator and nodded his head, "No, that is impossible. Big trees need big amounts of water." He had learned the art of irrigation from the Punjab, where flood irrigation was begun fifty centuries ago and has long been the only method known.

Three years later I visited Baluchistan for the fourth time, having helped to introduce drip irrigation into that region. Again, I found the same old farmer. This time he was sitting next to his well, sipping sweet tea, while his son was operating the drip system that now irrigates the same orchard. The old man was philosophical about the change. When I asked him if he now believed that big trees can be watered drop by drop, he shrugged and said, "By God, science can really do everything."

That old farmer in Baluchistan reminded me of the skeptics who dismissed the possibility of drip irrigation when it was first developed in Israel during the early 1960s. Although the idea itself was not new, it seemed to many experts that it could never become practical. Maybe it will work for flowerpots and in greenhouses, those irrigation experts said, but not for crops in the open field and certainly not for orchard trees.

Drip irrigation was indeed a revolution. It seemed to contradict common experience. According to the prevailing wisdom, a good irrigation regime was one that made maximum use of soil moisture storage. The main concern of irrigation researchers was how to define the effective lower limit of soil moisture storage. They concentrated their research on the dry end of the moisture range, in an effort to maximize the period between successive irrigations. To what value of dryness could a crop be allowed to deplete soil moisture without experiencing loss of yield? The search for that elusive limit was, incidentally, based on a misapplication of statistics. The failure to prove a significant difference in yield between a drier and a wetter irrigation regime was taken to be proof that no difference exists, hence the drier regime must be as good as the wetter. Crude experimental and sampling techniques or spatial variability often obscured the real differences. For a long time, preoccupation with the dry range of soil moisture prevented researchers from discovering the potential advantages of a moist regime--one based on the slow, frequent (or even continuous) application of water directly to the root zone and on the continuous maintenance of that zone in a nearly optimal condition of moisture and aeration. That realization finally dawned only in the early 1960s.

Drip irrigation did not appear suddenly as a full-fledged bright idea that immediately and automatically became a practical system. In fact, the idea itself was not new. What finally induced its evolution into a practical system was a combination of several synergistic factors. One was the advent of low-cost and durable plastic materials suitable for tubing and emitter fittings. Other important factors were the high cost of water and energy in Israel, as well as the need to overcome problems resulting from soil and water salinity, marginal lands and poor soils, expensive labor, and lack of security in outlying areas (necessitating automation). It was indeed a fortuitous confluence of conducive circumstances as well as of theoretical and technological developments that made drip irrigation an idea whose time has come.

Like many revolutionary advances, drip irrigation was at first heralded with exaggerated claims and unrealistic expectations. The justifiable enthusiasm for the new method carried certain dangers. In some cases, hasty adoption of drip irrigation without enough care in adaptation to local conditions resulted in disappointment. Drip irrigation offers many potential advantages, but it is no panacea. Inefficiency can occur as easily in the operation of a drip system as in the operation of conventional systems. This paper cites some of the potential advantages of drip irrigation and some of its limitations and remaining problems.

Advantages and Limitations

A primary advantage of drip irrigation is the possibility of obtaining favorable moisture conditions even in otherwise poor soils (such as gravels, coarse sands, and clays) which are ill-suited to conventional ways of irrigation. Another advantage is the capability of delivering water uniformly to plants in a field of variable elevation, slope, wind velocity and direction, soil texture, and infiltrability. The overall advantage is the ability to maintain the root zone at a highly moist yet unsaturated condition, so that the plants are never subjected to moisture stress. Soil air remains a continuous phase capable of exchanging gases with the atmosphere, thus ensuring proper aeration. High moisture reduces the soil's mechanical resistance to the penetration and free proliferation of roots. Where soil salinity is a hazard, the continuous supply of fresh water ensures that the osmotic pressure of the soil solution remains low near the water source. Moreover, as drip is applied beneath the plant canopy, it avoids the danger of leaf scorch resulting from the evaporation of brackish water, and it reduces the incidence of fungal diseases, both of which may occur under sprinkler irrigation. Since drip irrigation wets the soil only in the immediate periphery of each emitter, the greater part of the surface (particularly the area between rows) can remain dry and hence is less prone to infestation by weeds and to compaction by traffic. Direct evaporation from the soil surface is likewise reduced. In addition, pressure (and hence energy) requirements are decreased relative to those of most sprinkler systems.

The fact that drip irrigation wets only a fraction of the soil volume can also be a problem. While it is a proven fact that even large trees can grow in less than 30 percent of what is generally considered to be the normal root zone of a field (provided that enough water and nutrients are supplied within this restricted volume), crops become extremely sensitive and vulnerable to even a slight disruption of the irrigation regime. There is thus very little latitude for error or malfunction. If the system does not operate perfectly and continuously, crop failure can result rather quickly, because the reservoir of soil moisture available to the plants is very small. Other problems associated with drip irrigation are the accumulation of salts at the periphery of the wetted circle surrounding each emitter, which can hinder the growth of a subsequent crop, and the possibility that inexact calibration of water application can result in excessive through-flow and leaching directly under the drip emitters. In most cases such problems can be overcome by good management, that is, by tailoring the emitter spacing and optimizing the per-emitter discharge rate in relation to soil infiltrability and the lateral spread of the water, as well as to the variation of crop water requirements during the season. It must be emphasized, however, that optimization of all controllable variables can be very difficult in the face of so many uncontrollable (and not easily measurable) variables. Clogging of emitters is a frequently encountered problem, which can generally be obviated by proper filtration, acidification, and algacide treatment.

Accomplishments of Research

From a very modest, tentative, and initially rather simplistic approach, drip irrigation has grown in quality and versatility to become a highly sophisticated, technologically advanced system, or rather, set of systems. This has been a result of intensive engineering and agronomic research and development. Every aspect of the system has been improved, often beyond recognition. Numerous alternative emitter designs are now available, varying in flow pattern (laminar and/or turbulent) and discharge rate. Some emitters are designed to be pressure compensating, some to be less vulnerable to clogging, indeed there seems to be an emitter for each season and location. Unfortunately, the claims of the manufacturers and salesmen cannot always be validated. Nor is drip irrigation limited to the application of water in discrete drops. Continuous flow tubes (either perforated or porous) are available which act as line sources rather than as a series of point sources.

Equally impressive are the advances achieved in the design and manufacture of various types of ancillary devices, such as filters (media and screen types), and pressure regulators for use on sloping land. Metering valves, which are able to deliver pre-set quantities of water and then to shut off automatically while triggering the volume-controlled delivery of water to the next section of land, are available. Most advanced are the centralized programmable control systems, based on modern computer technology, which can regulate irrigation in response to meteorological and other dynamic variables. Innovations in both hardware and software are involved in achieving such systems of automation. Not the least of the developments concerns the injection of chemicals into the irrigation supply (a practice now called "chemigation"), most important being the injection of fertilizers ("fertigation"). Drip irrigation has in fact been adapted and applied to a growing number of different crops in different regions.

These are but a few of the accomplishments of research in the last two decades. Nevertheless, there remain many problems requiring additional research.

Problems Requiring Further Research

The two classical questions of conventional irrigation, both surface and sprinkler, have long been: when to irrigate, and with how much water? To the first question, drip irrigation can provide a simple answer: as frequently as practicable, even continuously. To the second question, the answer is more complex: an amount of water sufficient to meet the current needs of the crop (at its particular stage of development) and to prevent salt accumulation, without exceeding the soil's infiltrability. Merely delivering the amount of water the crop needs may not be good enough, if a part of that water escapes by runoff from the surface or by drainage below the rooting zone. The water applied must enter the soil where it is needed most and be distributed in a volume of soil sufficient to allow crop roots to extract water and nutrients in the requisite amounts. These are rather obvious statements of principle which, however, must be quantified, and here's the rub: exact quantification demands research.

Irrigation requirements vary in time as the season progresses, the weather changes, and the crop develops (typically, from the vegetative growth stage through the reproductive and fruiting stages to the senescing stage). To monitor these changes, and to respond to them dynamically, we need reliable, frequent or continuous, measurements. As yet, however, our sensing instruments, measuring methods, and data analysis methods are not completely satisfactory. Our monitoring methods are especially inadequate for the inherently non-uniform conditions of drip irrigation. Should we base our irrigation strategy on sensing soil moisture, and, if so, just where should we locate our sensors and how should we integrate their responses over the relevant volume? Alternatively, should we gear our irrigation rate directly to evapotranspiration, actual or potential? And in that case, how should we measure evapotranspiration realistically in partially wetted, heterogeneous fields, to which conventional measurements may not apply? Finally, should we adjust the irrigation rate in accordance with the water status of the crop, and if so, what are the most apt measurements (for example, leaf water potential or canopy temperature)?

These questions (which are, of course, easier to pose than to answer definitively) may not be critical in conventional irrigation practice, as the large volume of soil moisture storage serves to buffer the crop against fluctuations of weather. But in the case of drip irrigation the whole issue of crop-water relations becomes so much more sensitive owing to the heightened dependence of the crop on the operation of the system. Direct evaporation of soil moisture likewise depends on the fractional area of the wetted soil surface, as well as on the fractional area shaded by the crop. Hence the relation of evaporation to transpiration must be defined more specifically for drip irrigation, especially in the case of young and sparse stands of row crops or orchard trees.

We do not know enough about how different types of crops fare under partial-volume wetting. Is there a definable minimum or optimum rooting volume needed to sustain a given crop, and how might that volume depend on soil characteristics as well as on the nature of the crop? A commonly repeated rule-of-thumb is that one-third of the soil's volume should be wetted, as a minimum, but that is arbitrary. In principle, the aim should not be to maximize but to optimize root growth in relation to shoot growth, but at present we have no clear understanding of how this elusive optimum depends on the irrigation regime.

All these issues may seem to be theoretical or academic problems, but they have practical implications. We cannot really automate irrigation efficiently unless we can provide the central control unit with the necessary relevant information on crop-water relations. Without a fundamental understanding of the physical-physiological system, we can only rely on old-fashioned trial-and-error methods, which must be repeated in each case and cannot be generalized.

Perhaps the most glaring problem demanding attention arises ironically from our very success in developing the technology of drip irrigation to such a high level of mechanization. Have we let our fascination with high technology take control of our research, and have we, in consequence, turned away from the majority of the people in this hungry world who really need drip irrigation? I am referring, of course, to the special needs and circumstances of developing countries, especially in arid regions, for which drip irrigation would seem to hold important advantages: economy of water use, increased yields, energy savings (relative to sprinkler irrigation), adaptability to small-scale farming and to marginal lands, and so on. However, instead of simplifying and economizing irrigation and bringing its benefits quickly to the people of the poorer and less industrialized countries, we seem at times to be doing the opposite. Perhaps we should retrace our steps and seek to modify our systems to make them more widely and easily adaptable rather than more intricate and specialized. In the nonindustrial countries, the important attributes are low cost, simplicity of design and operation, reliability, longevity, few manufactured parts that must be imported, easy maintenance, and low energy requirements. Labor economy is less important.

A significant contribution made in the mid-1970s was the concept of bubbler irrigation. The proposed system retained the principle of closed-conduit delivery of water to the crop in plastic tubes and the possibility of establishing a high-frequency, low-volume irrigation regime. Unlike drip irrigation, however, the bubbler system discharges water from wide-orifice tubes rather than from capillary emitters. The pressure requirements are thereby reduced and the need for stringent filtration is obviated. At the same time, of course, a degree of control is sacrificed. The water bubbles out of the tubes into small basins surrounding each tree or group of plants, and the basins are ponded temporarily until the water thus delivered infiltrates. Bubbler irrigation is a promising intermediate-stage technology for many potential applications. Unfortunately, no commercial interest seems to promote bubbler irrigation, perhaps because it is a concept rather than a directly marketable product. It may not be as profitable to sell as some of the more capital-intensive systems.

It is important that scientists and engineers continue their current research and direct some of that effort to seeking methods that are less complex and expensive, and hence more suitable for adoption by poorer countries. Surely this, too, is a professional calling worthy of our best efforts.

Chapter 8

RESEARCH PRIORITIES TO IMPROVE IRRIGATION TECHNOLOGY

Marvin E. Jensen

Water is the major factor limiting quantity and stability of agricultural production in many regions of the world. Crop yield is linearly related to plant water stress and to water used in evapotranspiration when water limits plant growth. Irrigation minimizes plant water stress that is due to soil water deficits and, in arid areas, controls soil salinity. Soil water deficits occur when irrigation does not meet crop water demand. Soil salinity, though related to irrigation, is controlled by irrigation and drainage.

Basic principles and practices for crop production and irrigation are well established. When the gap between potential and actual crop yields is large, one of the main causes is the lack of application of these basic principles, a result of physical factors and complex biological, economic, and social relationships. The primary physical factor usually is inadequate control and management of the distribution of water in terms of time and quantity relative to the needs of the crop. Plant nutrition, plant density, and weed and insect control practices are often contributing factors. Cost-effective research to correct the gap between actual and potential production requires an assessment of the production system as an integral unit. When irrigation and drainage are the major limiting variables, effective adaptive research on irrigation and drainage should be planned and implemented.

Adaptive research, which is applying known principles to a production system under local conditions, identifies physical, biological, and economic problems that need further study. Equally important, such research provides essential data needed by engineers and managers for designing and operating productive irrigation projects.

Effective adaptive research on irrigation requires that a program plan be developed, that priorities be established, that an implementation plan for allocating available human and financial resources according to needs and priorities be set up, and that a stable research organization and work environment be established. In many cases, additional training of scientific and engineering staffs also is needed.

Developing a Program Plan

Irrigation and drainage research is almost always badly fragmented. There are many problems, however, that are common to all irrigation projects in developed and developing countries. No critical, world-wide assessment of on-going irrigation and

drainage research, research needs, and essential irrigation research infrastructure
presently exists; guidelines for establishing a research component for each new or
rehabilitation irrigation project should be developed to assist planners and borrowers.

Most effective research organizations have established procedures for planning and
implementing research to solve problems or to provide basic data needed by action
agencies and consultants. The planning steps taken usually are to identify the problem,
set priorities, and develop alternative approaches. Resources are then allocated to
implement plans and research projects. Identification of the problem related to irrigated
agriculture involves identifying and quantifying factors causing the gap between
potential and actual crop yields and identifying related factors that prevent full
realization of agricultural production under irrigation. Problem identification may
require an approach of diagnostic analysis involving system evaluations on a number of
fields or farms. Program leaders often are assigned the responsibility of problem
identification.

Establishing Research Priorities

Research priorities are established by research organizations and the clients or users
of research, who need to reach agreement on the problems that should be addressed.
Program leaders are key players in this process. Once established, the research
organizations determine the amount of available research resources to be allocated to
high-priority research needs, and research programs are established. Research leaders
develop research projects using one or more of the research approaches that have been
identified. When adequate progress has been made, or problems have been solved,
research resources are redirected to other remaining high-priority research problems by
program leaders and research administrators.

Research needs typically are shopping lists of topics on which research is needed.
Research priorities are research needs that have been placed in priority relative to the
following: the mission and expertise of the research organization involved; the available
physical, human, and financial resources for conducting the research; and the needs of
users of research who are faced with irrigation problems.

Research Priorities for Developing Countries

In April 1987, a colloquium on "Irrigation Research Needs in Third World Countries"
was held at Wallingford, England. After attempts were made to reduce the list of topics
that were proposed and develop short lists of higher-priority topics to form a post-
conference questionnaire, 43 topics remained.[1] The list of these topics was sent to the

----

1. See C. L. Abernethy and G. R. Pierce, "Research Needs in Third World Countries,' in the Colloquium Proceedings, Hydraulics Research, Wallingford, England, 1987.

participants after the colloquium for review and assessment, and 75 of the 120 participants responded to the questionnaire. None of the 43 topics was relegated to the "low importance" category, but the respondents suggested 75 more topics to be added. The responses from the various disciplines were quite similar. The top ten high-priority topics were: performance data; water scarcity; performance criteria; health; measuring devices; erosion and sedimentation; sustainability; crop breeding; efficiency of water use; and performance monitoring. Next on the list was reclamation of saline soil.

Research Priorities for the United States

The Research and Education Committee of the American Society of Civil Engineers (ASCE) Irrigation and Drainage Division recently completed its list of research needs for 1988.[2] The needs were grouped into five general categories. The categories and the total number of topics listed, but not in order of priority, were:

Topic	Number
Water conveyance and control	19
Plant water requirements	12
Water quality	13
Ground water	14
Weather modification	6

A similar Task Committee developed four high-priority research objectives in 1983.[3] The four broad objectives were: improved irrigation systems; management of soil and water resources; improved water storage and delivery systems; and more efficient drainage systems and practices.

Research Planning and Implementation

The first issue that will need to be addressed is whether all of the needed research can be conducted by a unit of the organization responsible for planning and operating a project. A new multidisciplinary unit may need to be established or arrangements made with one or more separate research organizations to establish interdisciplinary research teams. A major advantage of adaptive research is that many problems often can be solved by transferring successful experience gained in solving similar problems in other projects with similar crops and soils. When new crops and soils are involved, new problems can be expected, which may require more research and time to solve. Carefully designed field

----

2. American Society of Civil Engineers (ASCE), "Research Needs in Irrigation and Drainage," Journal of Irrigation and Drainage Engineering (in review, 1988).

3. American Society of Civil Engineers (ASCE), "Status of Irrigation and Drainage Research in the United States," Journal of Irrigation and Drainage Engineering, vol. 110, no. 1 (1983), pp. 55-74.

experiments and field trials on one or more pilot farms may be necessary to gain the knowledge needed. Adaptive irrigation research generally should not be considered as a one-time activity. Many adequate technologies exist today, but technologies are constantly changing. Adoption of new technologies usually requires collecting new data or establishing new coefficients for functional relationships.[4]

The organization needing research or quality-control data must determine if it has a fully capable and qualified, multidisciplinary internal research unit. If not, the options are to use an existing public research agency or university, form a new research unit within the organization, or contract needed research with one or more universities or private research institutions.

Developing an effective research unit within an operating organization requires time and resources. Long-term stability in funding and support are necessary to retain qualified staff and to equip, upgrade, and maintain existing research facilities adequately. The problem is more complex if the organization has not previously employed people trained in the disciplines needed. Generally, it will be more efficient to make arrangements with an established, experienced, and successful research organization than to establish a completely new internal research unit. Evaluations of existing research organizations should consider their missions, flexibility, overall research plans, and training programs for their research staff to enable them to use modern approaches and facilities.[5]

If one or more suitable research units do not exist, the first step in implementing a successful research program is to establish a stable, interdisciplinary research group. In many cases, existing research units may be combined to reduce management and staff costs. The mission and goals of such an organization must be clearly established along with an efficient management and operating structure.

Research to improve crop production involves solving multidisciplinary problems and filling information gaps under site-specific conditions. This information is needed to improve project management and performance in delivering water to meet crop needs and in removing excess water. Much of this research requires interdisciplinary research teams. Sometimes the policies and practices of research organizations conducting "irrigation" research discourage their scientists and engineers from becoming involved in interdisciplinary research. In Doorenbos' view, "the plant physiologist, the agrometeorologist, the agronomist, and the irrigation engineer must learn to speak the same language. There is no doubt that they need each other, can support each other, and can strengthen each other, in giving advice to the highly capital- and labor-intensive

----

4. See M. E. Jensen, "Research Requirements and Philosophy for Irrigation Projects." Paper presented at the International Seminar on Operation, Maintenance, and Management of Irrigation and Drainage Projects, Denver, CO, October 1986, sponsored by the American Water Foundation (Fort Collins, CO: Colorado Institute for Irrigation Management, Colorado State University, 1986).

5. See M. E. Jensen, "Research Requirements and Philosophy for Irrigation Projects" (1986).

irrigated agriculture."[6] The sociologist should be added to this list. Until we accept the fact that the overall objective is to improve "production systems" involving plants, soils, climate, institutions, and people, irrigation research will have limited effects on improving crop yields and production.

Developing a Research and Implementation Plan

One of the first priorities for a new research organization is to assess existing knowledge and develop a strategic plan for its operations. This plan may involve several research programs. It should identify what the organization is now, what it expects to be in the near future, and it should contain specific and attainable research objectives and viable research approaches. It also should identify staff training needs and develop a long-term staffing program by disciplines and management structure.

An irrigation project is a "crop production system." Effective irrigation research typically requires interdisciplinary teams. Many major irrigation projects need site-specific data to improve planning and design, and they need monitoring and evaluation data to improve operations. Each major irrigation project needs and should have an adaptive or operations research component to provide these data. At a minimum, organizations responsible for planning, designing, and managing irrigation projects should either have a research component or have an effective arrangement with an established research organization.[7]

Organizers of such research typically have several objectives. The first goal is to improve project planning and system design. Basic data have to be obtained to solve problems anticipated or encountered after construction that adversely affect operation, maintenance, and management of the main water distribution and drainage systems. These data are essential for maintaining or increasing the productivity that is normally expected of newly irrigated lands. The last objective is to minimize the risk that the irrigation project will be adversely affected by such problems as waterlogging and salinity. Specific objectives for a project research organization should be based on a diagnostic analysis and evaluation of system performance. Like private businesses that have large capital investments, many irrigation projects need a research component to remain competitive under changing world markets. An irrigation project should be run as if it were a business.

----

6. J. Doorenbos, "Water Consumption Related to Dry Matter Production: General Report," in A. Perrier and C. Riou, eds., Crop Water Requirements. International Commission on Irrigation and Drainage (ICID) International Conference, Paris, September 1984 (1985).

7. See M. E. Jensen, "Research Requirements and Philosophy for Irrigation Projects" (1986).

Financial Support for Research

Unfortunately, during periods of tight budgets and limited financial resources, research programs are often the first to be cut or reduced. Needed research requires top-level support. Key officials in an operating organization must be fully supportive of an adaptive or operations research program to improve project performance. Similarly, key policy makers must be supportive of the need to solve problems in order to assure adequate and stable resources for irrigation research. External support for a research component by international organizations like the World Bank, FAO, and USAID may be needed to assure implementation of a research program. This support could make the difference between success or failure of a project.

There is very limited support for irrigation research relative to that for more "sophisticated" research such as in medicine or basic sciences. This lack of support is of special concern in those countries where output from irrigated agriculture is the primary national product and where national production of food and fiber must be increased continually to meet the requirements of populations that are increasing at exponential rates.

In many countries, the gap between food production and requirements is widening. For example, Egypt now imports about half of its food supply, and its population is projected to double from its present 51 million in just 27 years.[8] India's population of 817 million is expected to double in 32 years. The momentum of population growth is staggering. Even in China, with stringent population control programs, the current population of 1.1 billion is projected to double in just 50 years. These trends indicate that greater support for irrigation-related research will be needed in many countries if production is to keep pace with population growth. In addition, increasing demand for high-value crops that require more precise water control will require accelerated adoption of improved irrigation technology.

Performance Monitoring

The lack of success of so many irrigation projects indicates that target objectives, along with a corresponding set of performance criteria or parameters, may not exist. Performance monitoring and evaluation is essential to identify components of the system that constrain crop production. Research may be needed to quantify cause-and-effect relationships. Such research may be called diagnostic analysis. Monitoring performance relative to established optimum and minimum standards is a first step in making improvements.

Where performance monitoring does not exist, especially in countries that depend on irrigation for production of their primary sources of food and fiber and where current

----

8. World Bank, The World Bank Atlas (Washington, D. C., 1987).

crop yields are much below potential yield levels, a research infrastructure of some type needs to be established. The overall objective of the research program would be to develop the technology needed to integrate modern agricultural production technology and the corresponding water requirements with the operation and management of irrigation and drainage systems. Full-scale pilot tests may be required to demonstrate this technology and the resulting benefits of such integration to project managers. These tests may be used to train both project personnel and farmers in irrigation water management. Technical support for project research could be provided by an international support group so that each project does not need a highly trained full complement of research staff.

Performance monitoring is associated with adaptive research. New corrective programs are often implemented without adequate knowledge and a clear understanding of current conditions and quantitative relationships between current conditions and resulting impacts. For example, there is a continuing large gap between design irrigation efficiencies and actual irrigation efficiencies because meaningful monitoring of project performance is rarely conducted. Knowledge of why project performance is below expectations is vital to the implementation of effective corrective measures.

In developed countries, occasional comprehensive performance monitoring is used to pinpoint areas that need improvement. For example, because of the need to reduce drainage volumes caused by deep percolation, the Westside Resource Conservation District in California conducted an irrigation performance evaluation on 83 fields in 1986-1987.[9] The results showed that nonuniform water application is the primary cause of deep percolation with all irrigation methods.

The purpose of irrigation is to eliminate, or minimize, the effects of inadequate soil water on plant growth. The purpose of an irrigation system is to make water available as needed by crops. The overall performance of an irrigation project must be based on its primary purpose--crop production. Monitoring the performance of components of an irrigation project and crop productivity is necessary to implement meaningful improvement programs. A viable research infrastructure with adequate support is a very important element in improving the performance and productivity of irrigation projects. If a research infrastructure exists, then jointly established research priorities should be the first priority, followed by planning and implementation of an irrigation research program.

----

9. C. M. Burt and K. Katen, "Westside Resource Conservation District 1986-87 Water Conservation and Drainage Reduction Program." Technical Report (March, 1988).

General Discussion

Mr. Hillel said he was intrigued with the straight-line relation between crop yields and evapotranspiration presented by Mr. Jensen but asked for suggestions on practical procedures to measure evapotranspiration. He also commented that there is concern that a stage is being reached where impressive gains in irrigation development are being offset by actual losses in irrigated areas through high water tables, salinity, and so on. He indicated that there are techniques to slow down this deterioration, which should be used to address this problem.

Mr. Jensen agreed that it was impractical actually to measure evapotranspiration as part of irrigation management. However, daily evapotranspiration can be estimated with plus or minus 10 percent accuracy, and provision of this information to farmers, for instance through daily radio broadcasts, could lead to lower costs and more efficient water application. Mr. Hillel suggested that the whole concept of efficiency of water use should be reexamined. For instance, it is difficult to distinguish between evapotranspiration and losses through seepage, and it would thus be possible to have efficient evapotranspiration even if irrigation efficiencies were low.

In commenting on research priorities, Mr. Burt indicated that the Bank should not be involved with the details of irrigation research but in its projects should focus on broader issues, such as constraints on efficient water delivery, operations, research and evaluation, communication between irrigation-system staff and farmers, and so on.

Mr. Hennessy drew attention again to the results of the recent seminar in England mentioned by Mr. Jensen, which was attended by thirty members from developing countries. He stressed the importance of the relevance and coordination of the research infrastructure, which in the United Kingdom is achieved though the Research Councils in conjunction with universities and institutes. It has been difficult to obtain cross-boundary coordination with this structure, however; there is now more emphasis on interdisciplinary aspects, but there is still a long way to go toward efficient, interdisciplinary research. With regard to research priorities, Mr. Hennessy stressed that we cannot do better until we know what we do wrong. We are not making effective use of the results of past efforts and seminars, such as the United Kingdom seminar, and this should be improved.

Mr. Fitzgerald stated that research should be broader than just engineering; it should include all physical, institutional, regulatory, and farmer-participation aspects. Institutional research is needed in particular because things tend to go wrong when governments get involved.

Mr. Plusquellec reminded the participants of the case studies undertaken in different countries by the Bank (about eight of them), which were discussed the previous day. These comparative studies, which for the time being are limited to gravity schemes and focus on the impact of design on irrigation efficiency and success, are expected to throw light on some of the issues discussed.

Mr. Manuellan expressed surprise that in enumerating the issues requiring solution, there was no mention of some of the irrigation technologies and research actions discussed the previous day, including those in his own presentation. There is a lot of technology known that could and should be applied. A successful project needs capable engineering and agricultural staff to plan and implement what needs to be done in the command area in terms of agronomy and irrigation design and water management. Donors need to satisfy themselves that technology is appropriate, that project management and institutional aspects are in place for project implementation, and that potential problems have been identified so that proper measures can be taken. This task is not the same for all projects because of variables due to location-specific conditions. What are the research activities in Bank projects?

Mr. Jensen agreed that some of the suggestions and proposed actions of the previous day's program had not been mentioned and wondered what kind of research the Bank would stress in the projects it finances.

Mr. Le Moigne indicated that the Bank is primarily a financing and development institution, which relies on the advice of other institutions and experts such as the workshop participants on what course to follow in research and technology development, transfer, and application, and seeking that advice was precisely the purpose of the workshop. Through its projects, the Bank finances many different types of research and is particularly concerned, for instance, that in irrigation schemes water availability is frequently not optimal in matching crop water requirements. Mr. Le Moigne agreed with Mr. Jensen on the need for interdisciplinary research. Increased farmer input is also desirable in deciding what should be adopted.

Mr. Khan said that in Pakistan there is little irrigation research as such. The Bank should insist on the inclusion of some adaptive research in all projects. Irrigation engineers as a rule do not coordinate with agronomists in delivering water, are not exposed to agriculture, and have no appreciation of sociological issues. For optimal use of available water, interdisciplinary coordination and farmer participation are necessary. Training at all levels and practical demonstrations, including field-drainage systems, are extremely important. But unless pilot schemes and demonstrations are appropriately designed, they will be less than effective. Governments should be sensitive to the needs of the people.

Responding to earlier questions as to the research activities included in Bank-supported projects, Mr. Barghouti explained that in the Gezira scheme in the Sudan applied research to improve water control and use of crop water is a major component. This research is conducted at the block level in the context of the existing irrigation system. Most of the Bank-financed research is carried out by departments of agriculture and irrigation.

Mr. Lahlou said that research should be carried out at different levels. Site-specific research based on location-specific characteristics is very important, but there is also research of a more general nature that could be carried out for groups of projects. National-level research would include the selection and development of different types of irrigation systems. Deficiencies in more fundamental research as well as field research are often due to lack of funds and support, which is a problem that needs to be addressed. Adequate research must also be accompanied by training. The dialogue between the engineer and the agronomist needs to be improved. Educational systems in some countries produce agricultural engineers who provide this link. Officials and bureaucrats also need practical training.

Chapter 9

ESTABLISHING RESEARCH PRIORITIES

Meir Ben-Meir

It is only twelve years to the year 2000. Research in irrigation alone can no longer meet the needs of modern agriculture. Such research, if approached by a single discipline, will not help prevent or even postpone the water crisis that several parts of the world will face by the year 2000. The need for a multidisciplinary approach can be illustrated by the events that led to salinization of the Aral Sea. This large lake, located 600 kilometers east of the Caspian Sea in the U. S. S. R., is fed by two large rivers, the Amu Dar'ya and the Syr Dar'ya. With the advent of modern irrigation practices, agricultural areas along the shores of both rivers increased considerably, using up the water that should have gone to the lake. In 1960, the Aral was the fourth largest lake in the world. Today, its 68,000 square kilometers have shrunk to 41,000 square kilometers; its water has been reduced by 40 percent, and, worst of all, it has turned saline. Its degree of salinity almost approaches that of sea water, and it is still increasing. Scientists have warned for years of an ecological disaster should the process continue uncontrolled. Twenty species of fish are now extinct, fishing has been eliminated, and the fertile land around the lake has turned to desert. The salt destroyed all plant life, including the cotton which had been the main crop in the region. In another twenty to thirty years, the Aral Sea will become another Dead Sea. This situation is only one of many that might have been prevented by adopting a multidisciplinary team approach.

The danger of increased specialization, of confining the problems to be solved within a single discipline, is that it often diverts us from the main goal of research and development itself. Before targets can be set, a prior assessment of the prevailing conditions--the infrastructure, environment, economy, and social factors--needs to be made. There is no model that can be applicable worldwide. But a general concept of target-setting can be conveyed, and Israel's experience is a useful model in this regard.

Historical circumstances in Israel created a unique situation, as two factors were brought together: most of the agricultural population had no previous agricultural background; and over 50 percent of Israel's land area is desert. These factors, taken together with the political, social, and economic need to make the desert bloom, form an almost sure recipe against conservative thinking, whether in research or development.

Israel's target is to continue to populate the arid regions as much as possible. The industrial and service sectors cannot yet take over the fulfillment of this national goal, and thus agriculture is the means used to increase population in the desert areas. Agriculture in the Western world is characterized by land consolidation, larger farms, and fewer farmers. Desert conditions dictate a different approach: the small farm and larger rural population. To sustain these small farms, engineers and scientists have to develop sophisticated technology. The Israeli model cannot be applied blindly to different geographic, social, economic, and political situations, however. This is an important factor that affects our technical assistance activities in developing countries.

Another important factor to keep in mind in learning from the Israeli experience is the ecological variability. Geographically, Israel is very small. It is just one day's travel by car from the northern snowy area to the southern warm region, with widely different climatic and agronomic conditions in between. Such an ecological setting calls for a dispersion of rural settlements throughout the various regions and highly diverse agricultural practices. What this situation suggests for other countries is that exact copying of other countries' models, without adaptation, means failure and that geo-political, economic, ecological, and social conditions define the national goals and determine research targets.

Even a focus on agriculture alone requires consideration of a wide array of environmental conditions to set targets for research and development, and the approach has to be broader than just research in irrigation. The development of drip irrigation systems is a useful illustration of how priorities were set in research and development activities in Israel. This technical development was not the result of a single discipline's approach; indeed, the potential in a multidisciplinary framework is almost unlimited. This paper reviews the development of drip irrigation in Israel, the problems it solved, the interrelationship between drip irrigation and other infrastructural factors, and its influence on the future of agriculture.

Half of the land of Israel is desert, and a great part of the water resources in this area is saline, with a salinity content of up to 1,600 milligrams of chlorides per liter of water, and sometimes more. When it was decided to develop settlements in the desert, under conditions of limited and low-quality water, an appropriate technology was needed to enable crop production under such severe conditions. The typical pattern of water distribution in the soil made it possible to use the drip system and develop favorable conditions for plant growth by maintaining relatively low salt content in the soil and leaching the salts outside the wetted zone. As a result, the roots can grow under relatively low soil salinity. This was the beginning. Over the years, it was found that drip irrigation has many more advantages.

In sprinkler irrigation, the common practice is to have long intervals between irrigations and to apply large quantities of water. As a result, plants suffer two types of stress conditions: poor soil aeration because of too much water during the irrigation itself; and, because of the long interval between irrigations, the plant is on the verge of wilting before it receives water. The plant enjoys optimal moisture for only a very short time between the two water applications. The drip system has a second very important advantage in making it possible to have a continuous water supply to the plant, as well as good aeration within the root zone. Development of devices, such as pulse systems, made it possible to give frequent water applications according to the needs of the plant and also increase the farmer's control over the operation of the system.

One of the most important advantages of the drip system was found in marginal soils. Soil surveys done recently with infrared aerial photography revealed that a great part of Israel's soils is shallow and another part has a poor infiltration rate. Under such conditions, frequent irrigation with sprinklers caused considerable water losses, mostly through evaporation and runoff, and despite continued irrigation, the plants suffered stress conditions. By wetting only a limited area of the soil surface, drip irrigation proved to be the only practical and efficient way to prevent water losses. In addition, marginal soils could be cultivated.

The next step was to integrate all the nutrients into the drip system. Use of computers helped achieve control over the combined irrigation and fertilization of the plant. Israeli farmers call it "feeding the plant with a teaspoon."

As a result of the accumulated experience and the fast feedback from the field to the research centers, the scope of irrigation has widened to include additional subjects. For example, sometimes there is an imbalance between the amount of energy the plant uses for foliage and fruit and the amount that goes into developing the root system. A root system that exceeds the optimal needs of the plant uses up energy needed to develop the fruit and foliage, thereby causing an imbalance and affecting the crop potential. Drip irrigation was found to be a useful tool in controlling the development of the root system by limiting the wetted volume in the soil. The result was a pot-like environment, even for field crops.

Another problem solved by use of drip irrigation is quite the opposite of the previous one. For mostly technical and economic reasons, a dwarf fruit tree was sought. The biochemical methods used were discarded with advent of the drip irrigation system. Citrus orchards with dwarf trees were developed, and because they were densely planted, harvesting expenses were considerably reduced. The development of harvesting machines for small-size orchards might be an easier target to reach than that of harvesting equipment for traditional orchards.

In addition to the need for a multidisciplinary approach in research and development, there is a wide range of considerations that decision makers in research policy are continually facing. The setting of research targets in irrigation and other water supply systems is based on several assumptions. First, the amount of water available for agriculture will decrease in the future. The world population and the standard of living will constantly grow, thereby increasing competition for the use of available water resources. Agriculture will thus be forced to use marginal water only-- marginal in quantity and in quality. The second assumption is that increased population creates increased pollution of water resources. Third, the cost of water will continue to rise because of the need to carry water from long distances and the expense involved in protecting water resources from environmental pollution. Fourth, agriculture will be forced in the future to use marginal and less-productive soils.

These assumptions rule out research that focuses only on sophisticated irrigation techniques. The wider scope of the relationship between irrigation and other factors affecting agricultural production--geographical, ecological, social, and economic conditions--has to be considered. In the past, the classic differentiation between extensive and intensive agriculture was characterized by the use of water only. Today it is not valid anymore. The increase in yields, due to increased use of water, does not justify the capital costs of bringing water to the arid regions. In the near future, it will be necessary to increase yields and income to such an extent that investments for new methods of developing water resources and their efficient utilization can be financed. Irrigation as a sole factor will not be able to carry this load.

In the not too distant future in Israel, the quantity of water will be limited, and it will be inferior in quality and expensive in price. Only a wide spectrum of research and development will enable the country to maintain intensive agriculture at the desired level. There are several areas in which the research efforts must be concentrated. Both chemical and biological means and methods for soil sanitation need to be developed as a precondition in the use of water. Methods and equipment for controlling environmental conditions need to be improved and perfected in order to save water and derive the maximum income from each unit of water. The relationships between water and other factors of production need to be tested in order to achieve the maximum benefit. All resources, including those of other disciplines need to focus on biotechnological research to develop plants resistant to disease and pollutants in soil and water and to breed high-yield, high-value crops so that the high cost of water can be financed.

Several instances in which such a coordinated, multidisciplinary approach is essential have already occurred in Israel. Citrus production is one of the main agricultural export sectors in Israel, but deteriorating water quality has endangered its existence. Research has focused on two main areas: to develop agro-technical means to create environmental conditions that will reduce the damage to the tree; and to use biotechnological methods to breed a citrus rootstock that is resistant to high salinity, which is a more radical and long-range approach. Scientists are at the midway point of this research project, and it will be at least another ten years until the results can be transferred into the field. This project has been a coordinated activity, which combines short-range agro-technical solutions with long-range biological ones.

Cotton is one of the largest intensive agricultural sectors in Israel. Cotton requires a great amount of water, and this sector will be endangered if the right solutions cannot be found. Israeli farmers have struggled to maintain cotton production by using drip irrigation systems, an approach that offers various benefits and represents a short-run agro-technical solution to the problem. But Israel faces an acute water shortage, which will encompass the whole region of the Middle East in the near future. Drastic steps in will encompass the whole region of the Middle East in the near future. Drastic steps in research are called for to achieve the goal of doubling output of the present crop without increasing the total amount of water used in irrigation. This ambitious aim cannot be achieved with research efforts of only a single discipline. Agro-technical and innovative methods must be combined to secure a reasonable chance of achieving the goal.

The most important factors that determine Israel's approach to its problems are water scarcity, a saturated local market, and the national goal of settling the desert. All these factors together have forced the country to adopt a concept that is different from that of the rest of the developed world, one that advocates development of the highly sophisticated, technologically advanced, intensive small farm that has control over a wide range of production factors. Such a farm, of which several models have already been developed in Israel, has to link levels of yield and income with units of soil and water if the farmer is to be able to withstand the limitations of scarce water and limited local markets. The production and income levels achieved in these few first models give reason to hope that agricultural research will succeed, even before the technological research, in finding a breakthrough that will increase farm income sufficiently to cover the high cost of desalinization of sea water.

From the Israeli experience, it can be concluded that although it is impossible to practice agriculture in arid regions without water, it is still a great mistake to try and concentrate all the research effort on irrigation alone. Other factors are no less important and have to be carefully considered. This concept, although successful in Israel, should not be copied as is, but the country's experience could be a pathfinder for those who can use it to their advantage.

General Discussion

Many participants complimented Mr. Ben-Meir on his talk. Mr. Simas agreed that irrigation research must cover many fields. This research entails a lot of surveys, monitoring, and evaluation of performance; research in on-going projects is extremely important. In Brazil, much of the research is conducted by private industry. Opportunities for gravity irrigation are limited. Pump schemes are therefore increasingly important, and energy considerations become paramount.

Mr. Zhang stressed the need for an interdisciplinary approach that moves from hardware to other considerations, particularly economics.

Mr. Sakena pointed out that whereas Israel is a young country, in older countries like India, research has to be geared to maintaining and improving old, established systems. He felt that extension and training are extremely important in the spread of technology and that research results are not always effectively used. It is also difficult to involve the users.

Mr. Goffin asked for confirmation that the cost of water development has to be covered by sectors other than irrigation. Mr. Ben-Meir confirmed that this was so and gave an example in which Israeli farmers previously had to be subsidized to export tomatoes. But development of new technology had resulted in a more favorable relationship between yield and water cost, and subsidies are no longer necessary.

Mr. Barghouti cautioned that the Israeli experience may not be replicable but also observed that the highly competitive world situation with respect to commodities necessitates research more than ever before to achieve greater efficiency and productivity and lower costs to enable further irrigation development. The question remains as to how one deals with this issue under conditions of more restricted resources.

Mr. Ben-Meir repeated that answers to those questions are location specific. Research can assist only with decision making based on local facts.

Mr. Le Moigne remarked that cotton research in Israel has focused on raising quality to increase farm income (maximum income per unit of water) rather than on raising yields (maximum yield per unit of water).

Mr. Hennessy asked Mr. Ben-Meir to give more information on the farmer in Israel and the marketing system for export crops.

Mr. Ben-Meir explained that exports are handled through one channel, aiming to provide maximum income for the farmer but allowing no competition by individual farmers. The system is not perfect but was found to be the best under the conditions that prevail. Farmers are organized in three sectors (two large and one small): community farms, farming cooperatives, and farms that are not cooperatives. About 90 percent of the land is owned by the government, which distributes land and water on a quota basis. Villages have about 100 families, and management is elected. Israel does not have surpluses because of the central control of production (mainly by quota).

Mr. Burt argued that many countries with Bank support could not afford research, and therefore they have to rely on transfer of technology from other countries. A lot of information exists, but what is needed is adaptive and operational research to determine the options.

Mr. Ben-Meir noted that both research and extension are needed to sustain development.

Mr. Hewron asked about the institutional decision-making process for setting research priorities.

Mr. Ben-Meir explained that there are two decision-making processes: researchers and farmers identify short-term needs at the farm level; and research and other government institutions set long-term priorities. Much of the research is carried out on farmers' land with their help.

Mr. Jensen related his experience with the process of setting research priorities in the United States, which also involves the users. These efforts mainly emphasize the short term, but government also reserves the right to decide long-term aspects that may carry high risk but also have high potential.

Mr. Barghouti asked about the decision-making process for allocating water to different uses under conditions of water scarcity.

Mr. Ben-Meir said that if water is scarce, the first need is a water act giving the government power to control the water. The government is to sustain a reasonable availability of water to meet the basic needs of the agricultural sector; over and above this level, water is made available at a higher price. In Israel, cities and towns treat their sewage to a standard for use in agriculture.

Mr. Hervé asked how one calculates the cost of urbanization and expansion of cities against the advantage of increasing agricultural output.

Mr. Ben-Meir responded that we must prevent people from moving to cities to less productive work but admitted that the activity-substitution question is very difficult.

Mr. Darghouth suggested that the Bank should include benefits of keeping people from moving into cities in its analyses.

In summing up, the chairman noted that the Israeli experience had some applicability to other settings and showed what could be achieved under conditions of marginality and scarcity, which provided a great challenge to develop innovative approaches. More innovation would be needed in the next generation to sustain development; problems need to be tackled on a broad scale using an interdisciplinary approach.

PART IV

TRANSFER OF TECHNOLOGY

Chapter 10

A CONSULTING FIRM'S EXPERIENCE

Salim W. Macksoud

To justify the recent interest in transfer of technology, one must assume three implied desirable qualifications. The transfer must be rapid, there must be minimum waste in resources, and the technology must be beneficial. Eventually all ideas spread around the world. Both beneficial and harmful innovations will be tried, and considerable losses might occur before the appropriate ones are adopted. To advocate and participate in the transfer of technology that has beneficial results, consultants must be sure of the suitability of the innovation. Unlike research centers, which test and evaluate various innovations, a consulting firm is asked to select an innovation and chart out the best means to adopt it, and it is thus held responsible for the suitability and efficiency of the innovation.

The Approach in Introducing New Technology

The first step is to know the technology. Recent innovations must be surveyed, and their applicabilities, problems, and requirements must be evaluated. Modifications might be required to make the innovation suitable to a particular situation. Implicit in this process is the second step--choosing a technology suited to specific conditions. In choosing an appropriate technology--whether an institutional innovation or a technical or engineering development--the consultant must be wary of partial, superficial, or "pseudo" technology, for which benefits are far outweighed by the cost of maintaining the innovation. If a scheme has to be maintained by imported equipment and spare parts and other operational inputs, the benefits cannot be long lasting no matter how spectacular short-term performance might be.

Consultants must rely on the results of recent research and on the records of similar attempts elsewhere under similar conditions. Valuable sources of such data for technical innovations are programs jointly run by national research centers and local dealers or manufacturers of equipment and instruments. For institutional innovations, the best source is the experience of the older irrigation authorities.

The selection process usually has to pass several tests before the innovation is accepted by the irrigation authority. Funding agencies often review plans with intense scrutiny, and some governments have recently required the consultant to be financially responsible for the engineering viability and integrity of the introduced innovation. Such attempts are to ensure that only proven innovations are recommended.

The third step is to introduce the technology, a task that usually requires less of the consultant than the more difficult process of selecting the innovation and testing its viability. It is usually the job of an irrigation authority to get farmers to make use of the new technology. Unfortunately, this is often easier said than done. The timing has to be appropriate for adoption of the technology, and the government agency has to carry out the required measures and activities that lead to successful introduction.

Factors Influencing Transfer of Technology

The classical demonstration approach--special pilot farms or individual farmer plots--seems to be the most successful method for spreading the use of an innovation. Of the factors that are important in adoption of new technology, this demonstration effort in itself is not the most significant. While the role of extension workers is important, water users will adopt an innovation if it is the proper one regardless of how poorly its introduction has been handled. The four factors that seem to play the major role in creating a favorable atmosphere and encouraging the adoption of new technology are the following: the psycho-economic mood of farmers; a suitable replacement technology; appropriate technical level; and sustained availability. All of these are important in determining if the proper environment exists for introducing new technology. When the "time" is right, the transfer takes place almost regardless of the manner of introducing the innovation. Anticipating the need before its "time" is right, even of a very effective innovation, does not seem to be a successful line of approach for introducing that innovation into operating schemes.

Farmers do respond to economic incentives. Farmers are usually traditionalists and, by instinct or nature, seem to resent change. But if they are continuously dissatisfied with their economic circumstances, they are responsive to change if it appears that the innovation will raise low incomes. This favorable mood among farmers can be triggered solely by a promise of exceptionally high financial rewards that are associated with the new innovation. In Saudi Arabia, for example, center pivots are the result of the subsidized price of wheat. As would be expected, when the price of wheat fell, the rate of technology expansion--even the number of operating units--declined.

An innovation that is economically and technically justified for a given circumstance is more readily adopted by farmers if it solves a problem that is perceived as the primary source of dissatisfaction. In other words, if the technology is a suitable replacement for the one single activity that farmers view negatively, they will be responsive to the innovation. For completely new schemes, the combination of factors that produces a proper time or environment for introducing a specific innovation need not include the dissatisfaction component, unless farmers from previously irrigated areas are being transferred to the new scheme.

The innovation has to be compatible with the level of technical knowledge, skills, resources, inputs, and support services available locally to water users. The level of technology cannot be out of sync with what is available or easily obtainable to make it work.

Farmers need to perceive that the technical innovation is sustainable--that it is here to stay. This permanence is maintained in various ways--through government authorities, private dealers and suppliers, availability of reasonably priced spare parts and replacement equipment--and such factors support the performance of the innovation.[1] Equally important is maintenance of the administrative and organizational aspects of the innovation.[2]

In summary, the four factors that influence the success or failure of an innovation are the following: dissatisfaction with the present situation leading to a will to accept a change; existence of an innovation that could profitably remedy the existing dissatisfaction; the compatibility of the innovation with the actual developmental level of the community or its adaptability to such a level; and the availability of a governmental or private source of support for sustaining the utilization of such innovation. These factors seem to work in unison to produce the atmosphere that favors the transfer of technology. They form, as it were, a four-link chain with which the transfer is pulled into position. The strength of the weakest link determines the pull that can be exerted toward the success of the attempt.

----

1. Examples of replacement equipment are special filter elements for combination pumping and filtering plants for drip systems in greenhouses.

2. Farmers need to be assured that various administrative tasks, such as preventing tampering with automatic controllers that regulate water delivery, assuring an equitable collection of charges based on volumetric readings of flow meters, or preventing the by-passing of such meters, are not neglected.

Chapter 11

THE ROLE OF EXTENSION IN IRRIGATION: THE ISRAELI EXPERIENCE

Moshe Sne

Development of water resources for irrigation is being carried out on a large scale all over the world. Huge amounts of capital are being invested in public water projects financed by national and international funds. In many such enterprises, construction of dams, pumping stations, reservoirs, conducting canals, and pipelines is of the highest level of technology. Nevertheless, poor utilization of water at the farm level is encountered, particularly in developing countries.

The question of optimal water use by farmers has worldwide importance. Water is a major limiting factor in food production in arid and semi-arid zones. In humid climates too, farmers irrigate crops to avoid losses during droughts and to maximize production. In some water-rich countries, shortage of funds may hinder full development of water resource potential. Improvement of water use can increase farm income, reduce occurrences of famine, and may rationalize investment capital.

The use of outdated methods of irrigation is one of the leading factors for poor water utilization, which contributes not only the waste of water but also to such negative results as runoff, waterlogging, soil salinity, soil compaction, leaching of nutrients, and inequitable water distribution.

Introduction of new irrigation methods can no doubt improve water use. Contemporary, modern irrigation technologies may provide solutions for almost any problem encountered. If farmers are convinced an innovation is profitable, have the money to invest in it, and are instructed in its proper use, they will willingly adopt new technology. A comprehensive system of funding, extension, and follow-up can enhance successful adoption of new technology. This paper reviews the Israeli experience in introducing new irrigation technology and the role of extension in both the domestic and international areas.

Establishment of the Extension Service

After the establishment of the State of Israel in 1940, hundreds of new settlements were established in the 1950s and 1960s. Many of the settlers were new immigrants who had no previous experience in agriculture. The conversion of merchants, artisans, or teachers into farmers required that an intensive extension system be built. Its structure somewhat resembled the Training and Visit (T & V) system. Quite soon after its establishment, it was apparent that the system did not meet the needs of extension in irrigation. A special service, the Irrigation and Soil Field Service (I&SFS) was established in the 1950s to deal exclusively with extension in irrigation. In 1970, the Service was awarded the Kaplan Prize, the highest award for productivity, for its indispensable contribution to efficient use and saving of water in the country. The award committee estimated the annual saving of water to be 150 million cubic meters, which represented more than 10 percent of the water used in agriculture. The Irrigation and Soil Field Service was later attached to the extension service of the Ministry of Agriculture, but it kept its uniqueness by sharing equally in the budget and responsibilities of the Farmers' Regional Councils.

Water is a national asset. It is the utmost interest of the government to ensure its efficient use. For farmers, water is a major limiting factor in production and income. Cooperation for better use of water is thus beneficial to both sides--a principle reflected in the joint efforts of the I&SFS, the Ministry of Agriculture, and the Regional Councils. The I&SFS is designated to serve farmers, and accordingly it has been organized into small regional, decentralized units, each of which has one to three field advisers. Each regional unit has a basic laboratory equipped with simple instruments. Budget difficulties eventually caused small units to merge into larger ones, but the principle of their having direct contact with farmers remained intact.

The regional units receive professional backing from the central unit, which has responsibility for coordinating professional activities, defining priorities at the national level, and providing regional units with up-to-date information. It serves several logistical functions, such as editing and printing publications, in-service training, and hiring of new field advisors. The central unit is staffed with highly qualified subject-matter specialists, who are kept well informed by the regional units.

Role of Extension Workers

Farmers who are knowledgeable and sophisticated tend to seek advisory help on their own initiative. Unfortunately, this is not the case with farmers who are less skilled; ironically, it is this group that badly needs help from extension workers. Most of the smallholders, new settlers, and Arab farmers fall into this category, and only efforts initiated by the extension service seem to yield results. In this task, leadership and skills in human relations are as important as technical expertise.

The extension worker is actually a missionary. Devotion and full identification with farmers are prerequisites to success. Confidence of the farmers is achieved by careful and gradual introduction of well proven technologies--ones that have been tried on a small-scale basis and compared with traditional methods. The most useful methods of introducing innovations have been demonstration plots, group meetings, field trips, and encouragement of a small core of cooperating farmers who serve as leaders to others.

The science of irrigation is complex and comprehensive. Because it includes soil science, hydraulics, operation and maintenance of irrigation equipment, engineering, mapping, climatology, and chemistry, proper extension in irrigation is beyond the capabilities of the ordinary generalist or field-crop advisor. The irrigation extension worker must have comprehensive expertise in these scientific topics, however; special attention has to be given to practical subjects. Technical skills and field experience are favored over formal degree.

Since no single advisor can hope to have extensive knowledge of all the topics that relate to irrigation, extension workers have to specialize to keep pace with innovations. In the regional units, each advisor specializes in certain topics, and information is shared with other colleagues. The most successful regional specialists are entitled to be part-time, country-wide, subject-matter specialists one or two days a week, a practice that enables them to be more professional, allows other regions to make use of their expertise, and prevents excessive growth of staff within the central unit.

Operation of irrigation systems must be expressed in quantitative terms. Water is measured in cubic meters (or gallons), intervals between water application in days, salt content in parts per million, and fertilizer in weight or volume units. This reality requires extension workers to be quantitative and to avoid vague recommendations, such as "apply more water" or "use less fertilizer." Instructions to farmers are given in explicit figures. Equipment that supports this quantitative approach are the soil auger (the "personal weapon" of the irrigation advisor), drying oven, pressure gauge, tensiometer, soil solution extractor, and field kit for determining electrical conductivity, pH, chlorides, nitrates, and nitrites. More advanced items are the full plant nutrition analysis kit, the neutron probe, and the pressure chamber.

Laboratory services are the ultimate support for the quantitative approach, and fifteen regional laboratories are operated all over the country. Farmer, field advisor, and laboratory are thus in close contact, and although the system is costly, the close proximity of the laboratory to the field enables a quick response to problems.

While irrigation is a major factor in agricultural production, it is only one of many. Success in agriculture depends on coordination of skilled agrotechnics, plant protection, irrigation, fertilizer, and farm management. For the extension service, this combination of factors means that coordinated teamwork is required at all levels from experts in the various specialties. When complicated problems arise, such teamwork yields the best results despite the time-consuming nature of the process.

Research and Extension

From the very beginning, the I&SFS did not view its role as merely a transmitter of know-how. When solutions were needed to crucial problems, it set up field trials and observation plots. About one-third of its effort is devoted to research. Each new innovation is tested in the field before it is supplied to farmers. Through close contact with manufacturers of irrigation equipment, staff members keep up to date. These contacts are easy and convenient to maintain because most of the irrigation equipment is produced in factories in agricultural settlements.

Cooperation with researchers is essential both for keeping extension workers up to date and for conveying the problems that farmers encounter. Fortunately, these contacts are very close, as most extension workers are eager to contribute to applied research as well as implement their ideas in the field. Most of the field trials are carried out jointly with researchers; joint professional teams, comprised of researchers and extension workers, discuss ideas and problems, coordinate experimental work, and formulate recommendations to farmers.

It is the conventional wisdom that developing countries are not mature enough to absorb high-technology innovations. The Israeli experience, however, makes this concept old-fashioned. Advances in electronics, hydraulics, and quality materials have made modern regulation and irrigation control devices that are cheap, reliable, and simple to operate. Careful and gradual introduction of automatic valves, pressure regulators, fertilizer systems, timers, and irrigation computers have brought about impressive changes in efficiency of water use and in crop yields. From 1951 to 1986, Israel's irrigated area increased fivefold, but water use increased threefold. Water consumption per hectare dropped, on average, from 8,500 to 6,000 cubic meters, and the output index from irrigated land, in fixed prices, grew tenfold. Productivity per unit of irrigated area doubled, while productivity per cubic meter of water tripled. At the national and regional levels, priorities are reassessed periodically. If a crucial problem is identified, a concentrated effort is made to remedy the situation as quickly and completely as possible, even at the expense of other activities if necessary.

Costs of the Extension Service

It is quite an impossible task to measure the benefit of extension, but it is a subject often discussed among farmer representatives. I&SFS employs 25 percent of the staff of the extension service, and the costs are relatively high. In addition to forty field advisors, an equal number of laboratory technicians operate the fifteen regional laboratories.

The regional unit of the I&SFS in Raanana District has four field advisors and four laboratory workers. The annual budget in 1986-87 was $150,000, which was shared equally by farmers and the national government. Irrigated area in the district totals

22,000 hectares, annual use of water is 150 million cubic meters, and 4,000 farmers are served by the unit. Expenditures for extension can be broken down as follows: roughly $7 per irrigated hectare, 0.1 cents per cubic meter of water, or $20 a year in direct expense to each farmer. With farmers paying water charges of 12 cents per cubic meter, the $20 expense for extension service support is equivalent to water charges for 165 cubic meters, or about 1.2 percent of the average annual allocation of water per farmer in smallholder villages.

From the farmers' standpoint, water savings would thus need to amount to 1.2 percent to justify their expenses for extension. From the government's point of view, in terms of benefit-cost, the savings in water that the extension service can promote would need to be only 0.17 percent in annual water consumption for the service to pay for itself. The marginal cost of new water resource development is estimated to be 30 cents per cubic meter; the cost of developing 500,000 cubic meters, which is less than 0.33 percent of annual regional consumption, is equal to the annual budget of the regional I&SFS unit. This focus on water savings does not include other benefits provided by extension, such as raising yields and improving land productivity. Because of such considerations, farmers have sought to keep the decentralized structure of extension services, including regional laboratories, despite the lower cost that would be involved in maintaining a central system.

The independence of the extension service has been maintained so that farmers would continue to have confidence in the system. The goal has been to seek fruitful cooperation with other government agencies also concerned with efficient use of water, such as the Water Commission, but not to enmesh farmers in bureaucratic requirements, such as inspection, credit, and other paper work, which might result if the service were not run independently.

Activities of the Extension Service

The extension service has played an instrumental role in increasing production of tree and field crops, carrying out scientific studies to reduce the effects of salinity, and conserving water and promoting its efficient use, particularly among smallholders. Implicit in these tasks is the introduction of modern technology, which is also a role served in Israel's cooperation with developing countries. An important element in its activities has been staff training.

Increased Production of Tree and Field Crops

Cotton was introduced in Israel in the early 1950s at a time when most of the cotton in the world was irrigated by surface irrigation. Little was known about sprinkler irrigation of cotton. Because of the difficulties Israeli farmers had in growing cotton--plenty of water was applied, but maturation was delayed and autumn rains depressed yields--the I&SFS gave top priority to quick implementation of new irrigation schemes for cotton. Under the direction of the central unit, dozens of observation plots were run simultaneously throughout the country. In a few years, use of water in the new schemes was cut to amounts that represented 50 to 70 percent of the water used in older systems.

Similar improvements were made in citrus production. The field trials undertaken in ten citrus orchards in the 1960s led to modernization of irrigation and its adaptation to local conditions. In the 1970s, micro-sprinklers and fertigation were widely used, and a central field trial led to various findings, the most important of which was the feasibility of applying nitrogen in irrigation water. These results were immediately implemented in dozens of satellite field observations elsewhere, and new irrigation and fertilization regimes were adopted.

Salinity was a major problem in areas growing citrus and avocado that used ground water of the coastal strip and water from the Sea of Galilee. The I&SFS set up a country-wide salinity survey in the early 1960s to evaluate the effect of water and soil salinity on these crops. Hundreds of observation plots were operated by a team from the central unit in cooperation with the regional units, which monitored water quality, salt content in soil and plant leaves, yields, and fruit quality for sixteen years. Researchers were able to employ statistical techniques to define the threshold of damage to different root stocks and varieties of citrus and avocado, thus enabling proper choices for new plantings according to water quality. The second stage of evaluating the impact of salinity on these crops, in a network of field trials, is currently under way.

Efforts in Water Conservation and Increased Efficiency

While efficiency of water use was thought in the late 1960s to be quite satisfactory on large farms and collectives, the waste of water was enormous on family-run farms of smallholders. The I&SFS gave top priority to finding out the reasons and correcting the problem. It designated four *moshavim* (smallholder settlements) as model settlements and monitored water use for three years. The staff tested innovative methods and equipment for their contribution to more efficient water use, which included the following: discharge and pressure regimes, plot boundary irrigation, automatic valves and pressure regulators, drip irrigation, fertigation, and patterns of irrigation episodes.

It was concluded that family-run farms could use water efficiently, although it required a more complicated and time-consuming effort than that for collective farms. The four villages became demonstration centers to further the adoption of new methods.

The I&SFS continues to pay special attention to increasing the efficiency of water use among smallholders.

Efforts were also undertaken to improve water use efficiency throughout the country. The Water Commission initiated a large-scale project in 1970, which was financed by the World Bank, to improve the efficiency of water use. Farmers could obtain grants and long-term loans to shift to water-saving irrigation methods. The Joint Regional and Central Judgment Committee, comprised of commission officials and extension workers, was set up to review each new irrigation plan. The examination of an irrigation system's feasibility in terms of crop and soil conditions, the hydraulics, the efficiency of water use, the operating system, and the costs brought about new standards of planning and implementation of irrigation systems in Israel.

The I&SFS played a major role also in the emergency conditions of 1986 brought on by two consecutive years of drought and the dwindling of aquifers and surface reservoirs. The Water Commission imposed a 15 percent cut in water allocations for irrigation in the 1986 crop year in the coastal strip; it made smaller reductions in other areas. It was feared that a curtailment of that magnitude in the coastal strip would have seriously reduced production and farm incomes. Concerted efforts were undertaken to minimize the damage, and ad hoc emergency irrigation recommendations were issued. Any factor that would reduce water consumption was taken into account, such as changes in growing season, shifts to drip irrigation, strict weeding, applications of fertilizer in water, and placement of old plastic sheets on greenhouse roofs. Computers generated water-use schedules for individual farmers. The schedule was updated monthly throughout the season according to actual water use and climatic conditions. Water-supply associations were able to compare planned and actual consumption, and any deviation was immediately noted and adjusted. Groups were active in explaining the background and reasons for the restrictions and the need to save water. The Water Commission, through its parallel but independent activities, sent out inspection teams to prevent violations of its allocation rules. The outcome was quite successful; despite the substantial cut in water use, agricultural production declined only marginally.

Introduction of Modern Technology

The I&SFS was instrumental in introducing modern technology in the Fara Canal Irrigation Project, in which 1,800 hectares of land in the Jordan Valley were converted from traditional surface irrigation to modern drip irrigation. Water consumption was cut significantly, irrigated area was doubled, and yields were increased impressively without any increase in the amount of water used. The area was cultivated by Arab farmers, many of whom are illiterate. The successful outcome of this project contradicts the notion that poorly educated people cannot adopt modern technology.

There were several factors that contributed to the success of the project. Careful, preliminary analysis was carried out by the local experiment station to determine the

optimal combination of equipment, irrigation, and fertilization regimes. The
demonstration effect was important; the palpable success of neighboring Jewish
settlements already using modern methods was significant in Arab farmers' adoption of
drip irrigation. In another project undertaken later across the Jordan River, the Rohr
Canal Project, farmers abandoned their preliminary idea of using sprinkler irrigation and
shifted to drip irrigation instead. The equipment chosen for the Fara Canal Project was
uniform, reliable, and simple, thus reducing maintenance problems. The innovations were
gradually introduced; the first steps were small in scale and financed by the national
government. Farmers thus undertook only minimal risk. Professional extension workers,
both public and private, provided long-term and dedicated follow-up and continually
solved problems as they arose.

Cooperation with Developing Countries

From the time when Jewish agriculture was first restored, it was clear that only
modern advanced agriculture could induce people to settle barren land, a goal that was of
the highest priority in Israel. Many farmers, agronomists, and researchers went abroad to
study agriculture in developed countries. The exchange of information continues,
particularly between Israel and the United States, to whom the country is greatly indebted
for its rapid progress.

Since the early 1950s, Israel has been cooperating on a large scale with developing
countries in transmitting know-how and contributing to their development. This
cooperation has taken different forms. In some cases the Israeli Ministry of Agriculture
has worked with its counterpart in other countries. Research institutions have also
initiated bilateral projects, such as the Israeli-Egyptian Desert Project.

Commercial companies in Israel that export agricultural equipment have followed up
on sales abroad with extension services that help the user learn the appropriate operation
and maintenance of the equipment. Such practices enhance sales and help prevent
equipment failures. A public company, Tahal Consulting Engineers, has helped train
extension workers as part of its large-scale water resource and agricultural development
projects in Iran, Thailand, Brazil, Kenya, Peru, and Costa Rica. In addition, many
private Israeli consultants, some of whom are former employees of the extension service,
work throughout the world and contribute their know-how and experience to developing
countries.

Such concepts and patterns of international cooperation have been changing
continuously in accordance with structural and institutional changes that have occurred
in developing countries. In particular, the concept of cooperation has replaced that of
assistance, which is typically a bilateral effort. Cooperation among many institutions is a
much more complex and painstaking effort, but the benefits of connecting funding
sources with expertise permits large-scale projects to be implemented.

Within the developing countries themselves, a significant change has been the establishment of local experiment stations and extension services that are staffed by nationals who have graduated from Western universities and institutes and returned home. Having a nucleus of qualified professional manpower changes the nature of a cooperative relationship, particularly as it helps avoid waste of resources, conflicts, and confusion.

The cooperative efforts of Israel have resulted in impressive successes as well as painful setbacks, but many lessons have been learned. The professionals are continuously working to improve their methods and keep pace with changing circumstances. One of the worst mistakes is to copy precisely a successful idea that has worked in one place without adjusting it to local, site-specific conditions. Diligent preparatory work and careful consideration of such factors as water and soil properties, climate, type of crops, farmer attitudes and skills, traditions, politics, and institutional structure are required before a cooperative effort can be successful. The introduction of new technology requires an effective extension service, one that is suited to local needs and potential, to provide instruction and close follow-up. It frequently takes a year to build an efficient, professional extension service, and the time required has to be built into the development project in the planning stage.

Whether extension services are centrally controlled or are decentralized is a matter than has to be worked out in specific situations by local authorities. While a central, professional staff is often considered essential, some countries are saddled with fossilized agencies that are unable to cope with modern needs. In some cases, grower associations run extension services of their own.

While the structure of an extension service has long been controversial, it is increasingly apparent that there is a need for greater specialization. Modern irrigation practices, including fertigation and problems of soils and salinity, are too complicated for extension workers who have only general training. To realize the benefits of modern plant-breeding efforts and agrotechnical improvements and to increase yields, farmers need better information about water applications. Improper water application not only affects the current crop but also has an impact in the long run on erosion, soil salinity, soil compaction, and water tables. All over the world, elevated water tables have forced farmers to abandon irrigated land. Only professional experts in irrigation can provide guidance on balanced and appropriate irrigation.

Extension workers are to serve farmers, but in an era of scarce, dwindling, and deteriorating resources, they must broaden their outlook beyond the benefits of individual farmers and regard such resources and the environment as a matter of national concern. Modern irrigation methods, particularly drip irrigation, need to be explained carefully by extension workers to counteract the tendency of new users, ones previously used to surface irrigation, to apply too much water. This focus on irrigation requires greater specialization on the part of extension workers, who in the past have had to deal with many other important aspects of agriculture, such as sowing, weeding, pesticides, harvesting, and marketing, often to the neglect of irrigation.

In addition to its increasingly specialized nature, the extension effort has to be a long-term one. It is a mistake to introduce new technology with only an intensive, short-term effort. Many failures of modern technology have occurred because of improper maintenance of equipment. Only continuous follow-up can ensure long-term successful operation. How many extension workers this follow-up requires depends on several factors, of which the most important are the size of the gap between the old technology and the new and the educational level of farmers. In special intensive projects, a ratio of one advisor to 500 farmers may be appropriate. If the extension worker is a subject-matter specialist who augments the activity of regional workers, a ratio of one advisor to 5,000 farmers may be adequate. In relatively backward areas in which extension work is just being introduced, a regional center might support five to ten villages, and the ratio of advisors to farmers may range from 1:500 to 1:2,000. In exceptional cases in which irrigation advisors are the only workers in the area, the appropriate ratio in the initial stage would be one advisor for every 200 to 300 farmers.

It is difficult to measure the efficiency of extension services. Factors that contribute to efficiency are the workers' motivation, skills, and methodology and the organization of the system that backs up their efforts, particularly in training them for successful group activity. A large-scale, rapid transfer of information is possible through group activity, in which a cooperating nucleus of farmers (roughly 3 to 10 percent) is trained by extension workers. With the use of observation and demonstration plots in their fields, this nucleus of farmers transmits information to wider circles of farmers. One extension worker can manage about five to ten demonstration plots. Greater efficiency can be attained by fostering competition among extension workers by collecting comparative data on yields, water-use efficiency, and so on. As the level of literacy rises in rural areas, short, simple, and understandable leaflets can provide seasonal information and thus help the extension effort.

With improved marketing systems, commercial firms can play a larger role in the transfer of technology. Suppliers of irrigation equipment and fertilizer often provide information on the use of the products they sell. Farmers typically regard such practices favorably, but the lack of objectivity is a potential drawback to involving the private sector in extension activities. It is essential for the government extension service and commercial firms to build mutual understanding and thereby ward off untoward consequences of promoting particular equipment or products.

Staff Training

The field advisor is the backbone of any extension service. Israel has been active in several different areas of training, both within the country at the Center for International Agricultural Development Cooperation (CINADCO) and in Europe through cooperation with the European Economic Community (EEC).

CINADCO has two programs for training irrigation specialists: the International Course on Irrigation and Extension; and short, "portable" courses for extension workers and farmers, which are run by small teams of Israelis in various countries. The combination of the two approaches, the "double site," has proved beneficial both to Israelis and the recipient countries. Israelis working in various countries are able to adjust the material taught in the course at CINADCO to meet the needs of participants, and those from abroad who have attended this course have an opportunity to learn practical information from the Israeli experience.

The course combines both theory and practice. In the classroom, students are grounded in the theory and prepared for a final project--designing an irrigation system and water distribution and fertilization schedules for a whole farm. There is no intention that through such an exercise these extension workers would replace irrigation engineers, but planning irrigation networks and water distribution schedules has proved to be a highly efficient means for learning to understand concepts of hydraulics and water management. In the most recent course, participants used computers to help plan the systems, and the initial impression was that they helped participants to understand irrigation principles better.

Participants spend one week in an agricultural settlement, practicing field work in irrigation. Substantial time and effort are invested in fostering group activity capabilities. The participants practice planning and implementing field days and field trips, and they use extension aids and prepare written materials for mass circulation.

In the short, portable courses, the emphasis is also placed on completing a planning exercise and gaining practical experience. Courses have been run in Africa, Southeast Asia, and Latin America.

Involvement with the European Economic Community has led to the establishment of training programs for hundreds of extension workers in southern Europe. A joint pilot project is now in its fourth year in the Mezzogiorno region of Italy. If the project is successful, the project will be expanded to other regions of Italy and neighboring countries. The need for such training stemmed from the difficulties these areas have had with irrigation; despite the building of new projects, on-farm water use has been poor. Agencies in these countries were not able to meet the challenge of providing extension services on a large scale for modern irrigation.

The project began in 1984 with a meeting of John Scully, former Chief Agricultural Advisor of the EEC, with M. Boaz, head of I&SFS, and a visit to Italy to coordinate planning with the local authorities and choose the three pilot project areas, Cassina, Pescara, and Foggia. The preliminary stages of the project were agreed upon in joint seminars held in Israel and in Italy, attended by delegates from pilot-project areas and experts from the EEC and Israel. Later that year, joint survey teams comprised of Italians, EEC experts, and Israelis, stayed in the three pilot areas, and they prepared a report to the EEC on long-term reconstruction of the extension system and on ad hoc activity for training of existing manpower and newly recruited staff.

Courses were taught--one in each of the two countries--and there have been three follow-up visits by Israeli teams during the irrigation seasons, which served to ensure strict on-farm implementation. Researchers from the University of Bari were included in the course in Italy, and the successful outcome suggests that there will be further collaboration of research and extension staff.

From the point of view of providing extension services, conditions in the two countries are not the same, but there were some areas of amazing similarity. Although there was plenty of water, farmers faced many difficulties. Use of surplus water overloaded conducting pipelines and canals, reducing pressure in pipes and lowering the water level in canals. Fields at higher elevations and those at the tail end of the canals were subject to low pressure and insufficient discharge.

With efforts of extension workers, problems were resolved by implementing strict irrigation turns, significantly reducing water consumption, and introducing uniform and reliable irrigation equipment. Yields on irrigated land were significantly increased. On-farm testing of several automatic systems is currently being carried out; further measures to introduce sophisticated irrigation systems will await the outcome of the testing.

The Israeli approach to the transfer of technology is characterized by this multinational cooperation among countries, the EEC, and Israel, by the contribution a developed country can make to another, and the continuing follow-up activity. Specialized extension in irrigation seems to be an efficient means to transfer technology. But the structure of the extension system in one country cannot be copied to another; each system has to be tailored to local conditions. Cost-benefit considerations have to take into account the income levels of farmers and the appropriate use of national resources. When farmers lack sufficient know-how, it is the extension workers job to provide information. Emphasis must be placed on the practical aspects, on cooperation with related institutions and staff, and on meeting the needs of farmers.

Chapter 12

INSTITUTIONAL INNOVATIONS IN RESEARCH AND EXTENSION

Shawki Barghouti and John Hayward

Irrigated agriculture worldwide is today facing serious challenges in several different dimensions. These dimensions include engineering design and technology, the economic viability of systems, the involvement of public and private institutions and of beneficiaries in sustaining irrigation systems, linkages between irrigated and rainfed farming, environmental concerns and the sociological and equity considerations of irrigation. We shall touch briefly on these aspects from the point of view of research, extension, and training, which we stress are intimately related and should not be divorced in the perception of issues surrounding irrigated agriculture. It is clear, however, that all situations differ in the balance and severity of problems. To generalize is unwise, but we do so deliberately, in the hope of provoking discussion from a wide range of viewpoints from which we may all learn.

Issues in the Design and Development of Irrigation Systems

Much has been said about engineering design and irrigation technology at this seminar, but the focus needs to be on developing more efficient systems for managing water in regions where water is scarce. These systems will by their very nature be sophisticated and will demand more understanding of their use by engineers, agronomists, and farmers if they are to succeed. Research should not, therefore, strive for technical efficiency without considering the practical reality of introducing such systems into communities whose access to knowledge of efficient water use is limited. The true costs of organizing and training farmers in water management must be taken into account, just as the potential social and economic risks of system failure must be carefully evaluated.

The Economic Viability of Systems

A major concern facing irrigation is how to build irrigated farming systems flexible enough to respond to changing market signals. The economic viability of irrigated schemes designed to produce a specific grain crop or cash crops is being challenged by the decline in commodity prices in world markets. In many cases, farmers in such systems are rushing toward production of higher-value crops, often ignoring the attendant risks involved in managing new crops about which little is known, especially in controlling insects and diseases and in producing sustainable qualities and quantities for domestic and foreign markets. Increasing diversity and flexibility of such systems will have to be addressed through better integration of research to produce improved crop varieties, efficient techniques for water control and distribution in irrigation systems, and development of systems to ensure farmers receive up-to-date information on market requirements and prices.

In Southeast Asia, for example, the success of the Green Revolution coupled with the decline in commodity prices has reduced the profitability of irrigated agriculture. As a result, existing irrigation schemes have been neglected, and new irrigation projects are economically difficult to justify mostly because of their low return on investment, which is largely caused by rigid agricultural production systems and commitment to a limited range of crops.

In the Middle East and North Africa, irrigated agriculture is faced with the challenge of reducing its reliance on traditional export crops: cotton, sugar and tobacco, and increasing the production of high-value crops (vegetables, fruits, oilseeds, and fresh fodder) and feed grains for the livestock sector. This shift requires a serious, but gradual, adjustment in research strategies and priorities, in extension programs and activities, and in the involvement of the private sector in research and extension initiatives.

In Africa, repeated occurences of drought, sensitive political dimensions of food security, and unreliable crop production from the rainfed agricultural sector have increased the pressure on irrigated agriculture to move away from producing cash crops, the main source of foreign exchange, to producing food crops urgently needed by the growing urban communities. The concern over the economic viability of such schemes is constraining donor involvement.

The Role of the Public and Private Sectors

Facing the challenges of irrigation today should not be the responsibility of the public irrigation development agencies alone. Recent studies by the World Bank indicate that increased involvement of the private sector in irrigated agriculture has indeed resulted in higher efficiency and profitability. The Jordan water resources sector study, for example, concludes that rapid and successful expansion of irrigated agriculture was enhanced by an aggressive private sector, which encouraged farmers to adopt drip irrigation, construct plastic houses and tunnels, and to use modern inputs and varieties to maintain a viable mix of high-value crops.

The perspective of the private sector tends to be short term and profit motivated. This leads to greater sensitivity to market opportunities and to the immediate advantages of improvements in technology. The private sector can react more rapidly to changing circumstances and is not constrained by bureaucratic procedures. Expertise and equipment can be procured, policies can be changed in response to changing circumstances, and incentives can be offered at all levels to promote efficiency. These advantages must, however, be balanced against the need for long-term environmental safeguards, against the possibility of resources being exploited, and the need for socioeconomic equity.

The private sector is efficient in dealing with advanced farmers who have high potential earnings and who need to be serviced almost on an individual basis. The private sector often supplies technical knowledge, inputs, market information, and market opportunities to these farmers. But monitoring such systems against the likelihood of pesticide misuse, for example, is a function of an objective public service. Similarly, as irrigation spreads to marginal farmers, the attraction of such schemes falls off, and, for sociopolitical and equity reasons, the public sector must step in to provide services.

Increasingly, the private sector has benefited directly from providing and installing more advanced irrigation systems. In circumstances where there is a longer-term, private-sector involvement, training in use of equipment and assistance in overcoming operational problems is provided by suppliers. Similar requirements must be made of the private sector even where government extension services are providing long-term support to marginal farmers. The role of the private sector in providing subject-matter specialists should be fully exploited by public services. Furthermore, the private companies should be made to realize that it is in their long-term interests to provide maintenance contracts and training courses.

Public and private sectors do not have separate and distinct roles; they are mutually reinforcing--in research, extension, and training. It is important that the roles of each be fully realized and exploited in the design and development of irrigation systems. The challenge is how to maintain a balance between profit motivation and socially acceptable and sustainable irrigation systems, especially in more marginal schemes.

Linkages between Irrigated and Rainfed Agriculture

In many countries, irrigated and rainfed agriculture forms a continuum involving the same farmers and the same crops. Research, extension, and socioeconomic factors should, therefore, be concerned with the total farming system, as changes to one aspect will obviously have an impact throughout the system. Restricting research programs within narrowly defined agroclimatic zones may reduce the flexibility of diversification along this continuum and will add unnecessary expenses and duplication of efforts. Research and extension are often compartmentalized not only between irrigation and rainfed departments but also between agronomy and engineering. It is essential to realize that much of the success of irrigation depends upon the effective management of the interface between crop and water. It is, therefore, impossible to divorce water delivery from crop uptake, and the engineering and biological dimensions of the system must be integrated in knowledge generation and dissemination systems.

The Human and Sociological Aspects of Irrigation

The impact of irrigation systems on human and physical environments both within and outside the irrigated area is particularly complex. It involves developing reliable methods to assess the long-term effects of irrigation and its impact on future generations of people and local ecosystems and at the same time maintaining a profitable performance of irrigated agriculture.

The role of farmers in managing, operating, and upgrading irrigation systems needs to be addressed throughout the production process. This requires that not only factors directly affected by irrigation, such as settlements and resettlements, beneficiaries and procedures be studied but also those who may affect the performance of the system, such as upstream communities of farmers and operators in water catchment areas. The World Bank work in this dimension is worth mentioning, especially its involvement in water users associations, the sociological aspects of developing and managing water catchments, and social forestry programs in the highlands of Ethiopia, India, Nepal, and several other countries.

A joint research effort is also required on catchment management, soil erosion, flood protection, and forestation. The social dimension must be an integral part of irrigation research, especially where it deals with soil erosion and reservoir sedimentation. The extent of participation of upstream farmers in soil management activities may have significant impact on downstream irrigated farming. Engineers, agronomists, and sociologists will be needed to study such effects and the means to mitigate them. National research systems in Pakistan and India are currently supporting multidisciplinary research inquiries into the relationship between upstream management on downstream irrigation performance. This effort is being augmented by increasing funds for studying and establishing water users associations. Social research in water management has received priority attention in the research program of the International Irrigation Management Institute (IIMI).

One important sociopolitical issue in the human dimension is that irrigation by its nature constrains the degree of freedom of individuals. Just as it would be totally inadmissible for road users to behave independently of the rules of the road, so it should be equally inadmissible for irrigation users to behave independently of the rules of the system. Accommodation of planting dates, cropping patterns, and crop-protection programs contributes largely to the common good. Some individuals, however, wish to adopt radically different systems. The development of water users associations has gone a long way towards promoting socially cohesive behavior in irrigation, but research, extension, and training schemes should be geared to promoting collective action and responsibility.

A related issue is the need for research on health and environmental aspects of water reservoirs, which experience with the High Dam in Egypt has made apparent. The evolving research strategy on such issues deserves special attention.

Specific Research and Extension Issues

We wish to focus briefly on some specific issues of research and extension that relate to the general topics just discussed.

Research Links to Crop Production, System Design, and Water Delivery

At issue is whether irrigation and hydraulics research should be institutionally integrated with research on agriculture and crop improvement, or whether each type of research should be carried out separately. The reality of the situation calls for separate efforts. In most countries, irrigation and agriculture departments operate separately, each with its own program of work and budget. Any attempt to amalgamate institutions into one bureaucratic unit is likely to fail. A system has to be established whereby these agencies jointly study the feasibility of upgrading, modernizing, or modifying existing irrigation systems for cultivation of crop mixes. Such studies should go beyond engineering design and hydraulics research. They should cover crop performance, the adequacy of the marketing systems and transportation, and the potential for regional specialization within irrigated schemes and between irrigated and rainfed crops. Such a model has been successfully demonstrated by the International Rice Research Institute, especially its recent work on agricultural diversification in Asia.

The economic feasibility of upgrading irrigation systems needs to be studied. Costs of improving drainage and water control for existing irrigation systems, building storage capacity, and lining water channels have to be compared with the value of crops grown on the land. This is a matter of market demand. Any study of upgraded irrigation systems has to consider other investments in infrastructure, including roads and communications, and in marketing, particularly in packaging, drying and storage facilities, trucks, and refrigeration, which may be complementary. Hence, any applied research for irrigation development requires the efforts of multidisciplinary teams of researchers dealing with engineering, agronomic, socioeconomic, and institutional factors of irrigated agriculture, regardless of the bureaucratic boundaries separating those research organizations. The establishment of multidisciplinary research projects is likely to overcome bureaucratic hurdles, improve the relevance of research issues, and increase the possibilities of dissemination and adoption of research results. The national agricultural research system in Indonesia is now being designed to address these requirements in different regions of the country.

Water Requirements for Crops

Irrigated areas provide special opportunities and special problems. With favorable temperatures and good drainage, irrigated areas should theoretically be suitable for a wide range of crops. Unfortunately, however, many irrigation systems were designed for specific crop production and represent a physical commitment to that crop's water requirements. Building flexibility into irrigation systems to allow switching among crops is a special engineering problem, which deserves attention in conjunction with research on crop and water requirements. In the Gezira and New Halfa schemes in the Sudan farmers were producing cotton with only six irrigations when all research recommendations called for twelve irrigations. Cotton yields were consequently low, and farm income from irrigation declined. Ideally, farmers would have reduced the cotton area or shifted to other crops that used available water more efficiently. By doing so, they would have saved on labor, seeds, and other inputs. The absence of research on these options curtailed advancement in sustaining cotton production and reduced the farmers' potential incomes.

Crop and Livestock Production

An analysis of the comparative advantage of crop and livestock alternatives in irrigated regions is needed. Agricultural and physical scientists should collaborate with economists and sociologists on this problem. The focus should be on those commodities that promise higher incomes for farmers and that utilize water more efficiently. The increasing demand for high-value crops and for meat in the growing urban areas of North Africa and the Near East has moved the issue of livestock production in irrigated areas of these countries to the forefront. Research is needed to assess the trade-off between crop and livestock production under irrigation and the consequent implications for irrigation design and water requirements.

An important aspect of such research is that new production technology can significantly alter a country's underlying comparative advantage. Because of this dynamic aspect of comparative advantage, its study needs to be a continuous process, not a once-and-for-all exercise.

Appropriate Extension Systems for Irrigation

Concern has recently been expressed that the didactic Training and Visit (T & V) approach to extension may not be suitable for irrigation areas. This criticism stems from two concerns: that irrigation farming involves more complex information than the simple messages that are regularly disseminated through most T & V systems; and that group rather than individual behavior is essential for irrigation areas. Such criticism fails to recognize the main rationale for T & V, which is to bring monitorable, structured order to an extension service, thus enabling it to interact efficiently with farmers on their farms.

There is no reason whatsoever why extension workers, operating according to T & V schedules, should not interact with groups.

In Pakistan, for example, the water users associations are ideal centers of focus for the extension service, and a decision has been made to ensure that each chairman of an association is by definition a contact farmer. This should bridge the former gap whereby the Water Use Authority (WUA), under the On-Farm Management Department of the Ministry of Agriculture, was unrelated to the extension service, which was administered by a separate department.

The complexity of messages should not be a barrier to the introduction of structured extension. The front line agricultural extension worker cannot be expected to be a master of all technical trades. It is, however, important that extension workers are supported by subject-matter specialists who can answer farmers' problems when these are brought up at training meetings. The skill expected of an extension worker is not highly technical; it must be first and foremost interpretive and advisory. Awareness of sources of appropriate knowledge is as important to the extension worker as having a broad knowledge of farming problems.

Whatever the system of extension employed, it is critical that farmers are met regularly on their own, or representative, farms and that there is a two-way flow of information--from public or private research and from markets to farmers, and from farmers back to research. Furthermore, extension, supported by technical specialists, should focus on the entire farming enterprise, rainfed and irrigated, and not on any specific technique such as water delivery or crop protection.

A question frequently asked is what is the appropriate ratio of extension workers to farmers in irrigation schemes? There can be no correct answer to this question. Just as irrigation systems are designed to fit the circumstances, knowledge transfer systems must be designed to be flexible, to fit the increasing sophistication and educational levels of farmers, to relate to the technical needs of commodity campaigns, and to suit seasonal requirements. Similarly, if, as in Taiwan and Thailand, the private sector is strong or middlemen in the marketing chain are important purveyors of technical knowledge, it is reasonable to reduce the extension input.

It is more efficient to appoint the best qualified and most effective extension staff to irrigation areas, because higher-value crops require more careful nurturing and more precise timing of operations to achieve their high potential. Irrigation farmers therefore need a more rapid response to their problems, which favors the private contracted advisor or the dedicated vigorous extension worker. In time, of course, farmers themselves gain knowledge and confidence to interact directly with research or the private sector, at which point the intensity of extension input can be reduced.

In conclusion to this brief report, it should be emphasized that water, crops, and farmers can no longer be studied separately. Engineering aspects of water delivery must be linked to the crops that are to be grown, to the capabilities and capacities of the farmers, to the comparative advantage of the irrigation area, and to the potential impact on health and the natural environment. Research on irrigation tends to focus on the search for new water-delivery technology. Yet there is no shortage of innovative ideas; some 7,000 of them languish at the moment with the U.S. Patent Office. The critical issue is that research on the relevance and adaptability of irrigation techniques is lacking.

Attention should be focused on the respective roles of the public and private sector in both research and extension; it is a blend of the two that is likely to produce the most immediate and the most sustainable results. The generation of knowledge and the requirements of disseminating irrigation technology are complex, but a structured extension service, interacting with farmers' groups and linked to subject-matter specialists from all appropriate disciplines, can effectively serve irrigation needs.

The appropriateness of innovations in irrigation research and extension for a particular country obviously depends upon the stage of development of that country. It may be that some of the countries represented at this symposium have passed the stage at which there are problems in bridging engineering and agronomy or in coordinating the interactions of the public and private sectors to best effect. But many countries have only begun to face the issues outlined here. We hope that our discussions will assist in developing their irrigation programs.

ANNEX
LIST OF PARTICIPANTS

WORLD BANK

WORKSHOP ON INNOVATIONS IN IRRIGATION

April 5-7, 1988, Room E-1244

List of Participants

Country Representation

BRAZIL
Mr. Jose R. Simas   Director General, PRONI

CHINA
Mr. Zhang Zezhen   Adviser, Ministry of Water Resources and Electric Power

EGYPT
Dr. Abu Zeid   Director, Irrigation & Drainage Center

FRANCE
Mr. G. Perrin de Brichambaut   Senior Technical Adviser, Ministry of Agriculture
Mr. G. Manuellan   "Ingenieur General", Ministry of Agriculture
Mr. J.J.Herve   "Ingenieur en Chef", Ministry of Agriculture

INDIA
Shri Kr Saxena   Central Board of Irrigation and Power
Shri Marouthi Babu   Central Board of Irrigation and Power

ISRAEL
Mrs. Yahalomah Shehory   Agricultural Attache, Embassy of Israel
Mr. Ben Meir   Director General, Ministry of Agriculture
Mr. David Melamed   Senior Irrigation Engineer TAHAL Consulting Firm
Mr. Moshe Sne   Extension Service, Ministry of Agriculture

JORDAN
Mr. Abdul E. Khatib   Embassy of Jordan
Dr. Munther Haddadin   Adviser to the Prime Minister
                                     Past President, Jordan Valley Authority
Mr. Salim Macksoud   Principal Partner
                                     Dar Al Handasah, Consulting Firm

MEXICO
Mr. Joaquin Huerta   Instituto Mexicano de Tecnologia del Agua

MOROCCO
Mr. Othmane Lahlou   ICID President and Director, ORMVAG Kenitra

PAKISTAN
Mr.Baz K. Khan   Director General
Federal Water Management Cell
Ministry of Food, Agriculture and Cooperatives

UNITED KINGDOM
Mr. W.R.Rangeley   Honorary President, ICID
World Bank Consultant

Mr. John Hennessy   Partner, Sir Alexander Gibb & Partners Consulting Firm
Vice President, ICID

UNITED STATES
Mr. C. Burt   Professor, Department of Agricultural Engineering
California Polytechnic State University San Luis Obispo

Mr. M. Jensen   Director, Colorado Institute for Irrigation Management, Past President ICID

Mr. J. Keller   Professor, Utah State University

AGENCIES

INTER-AMERICAN DEVELOPMENT BANK (IDB)
Messrs. A. Blandon, H. Suarez, L. Garcia

US AGENCY FOR INTERNATIONAL DEVELOPMENT (USAID)
Messrs. R. Backus, C. Martin   Africa Bureau
P. Novick, S. Peabody   Asia Bureau
J. Corvick, C.Anderson   Latin America Bureau
W. Fitzgerald   Science and Technology Bureau
P. Reiss   USAID Consultant

WORLD BANK STAFF

Mr. Paul Arlman   World Bank Executive
Director

AFRICA REGION
Messrs. Van Tuu Nguyen, A. Elahi, AFTAG; A. Khan, R. Hewson,
K.Loganathan, AF2AG; J. Weijenberg, AF3AG; K.Singh, R.Faruquee,

A. Seth, AF4AG; S. Darghouth, E. Garfield, D. Lallement, AF5AG
J. Ter Vrugt, AF6AG

ASIA REGION
Messrs. B. Albinson, H.Frederiksen, J. Berkhoff, T. Daves, Y. Choi, ASTAG;
van Voorthuizen, AS1AG; E. Moerema, I. Naor, J. Roman, AS2AG;
C. Diewald, C. Perry, AS4AG; D. Leeuwrik, AS5AG

EMENA
Messrs. W. van Tuijl, P. Streng, J. Guillot-Lageat, C. Cheng, EMETAG;
T.Yoon, H.S.Thavaraj, E.Roell, J. Mohamadi, B.Abbai, G. Motha,
EM1AG; J.M.Villaret, EM2AG; P.C.Garg, EM3AG

LATIN AMERICA REGION
Messrs. J.L.Ginnz, LA1AG; A. Cornejo, LA2AG; J.Heidebroek, J.Martinod,
LA3AG; P. Wittenberg, LA4AG; E. Idelovitch, LA41E; M. Mc.Garry,
LATAG;

SECTOR POLICY AND RESEARCH (PRE)
Mr. E. Tillier, VPPRE

Agriculture and Rural Development Department
V.J.Vyas, AGRDR

G.Le Moigne, H. Plusquellec, W. Ochs, U. Kuffner, J. Hayward, S. Barghouti,
T. Pritchard, AGRPS

A. Braverman, AGRAP

Infrastructure and Urban Development Department
S. Arlosoroff, INUWU

Environment Department
D. Hillel, ENVST, M. Ahmad, ENVOS

IFC
Engineering Department
Z. Oumer, CENAF

World Bank Consultants
Messrs. Paul Goffin, Syed S. Kirmani, E. Gazit

Distributors of World Bank Publications

ARGENTINA
Carlos Hirsch, SRL
Galeria Guemes
Florida 165, 4th Floor-Ofc. 453/465
1333 Buenos Aires

**AUSTRALIA, PAPUA NEW GUINEA,
FIJI, SOLOMON ISLANDS,
VANUATU, AND WESTERN SAMOA**
D.A. Books & Journals
11-13 Station Street
Mitcham 3132
Victoria

AUSTRIA
Gerold and Co.
Graben 31
A-1011 Wien

BAHRAIN
Bahrain Research and Consultancy
    Associates Ltd.
P.O. Box 22103
Manama Town 317

BANGLADESH
Micro Industries Development
    Assistance Society (MIDAS)
House 56, Road 7A
Dhanmondi R/Area
Dhaka 1209

BELGIUM
Publications des Nations Unies
Av. du Roi 202
1060 Brussels

BRAZIL
Publicacoes Tecnicas Internacionais
    Ltda.
Rua Peixoto Gomide, 209
01409 Sao Paulo, SP

CANADA
Le Diffuseur
C.P. 85, 1501B rue Ampère
Boucherville, Quebec
J4B 5E6

CHINA
China Financial & Economic Publishing
    House
8, Da Fo Si Dong Jie
Beijing

COLOMBIA
Enlace Ltda.
Apartado Aereo 34270
Bogota D.E.

COSTA RICA
Libreria Trejos
Calle 11-13
Av. Fernandez Guell
San Jose

COTE D'IVOIRE
Centre d'Edition et de Diffusion
    Africaines (CEDA)
04 B.P. 541
Abidjan 04 Plateau

CYPRUS
MEMRB Information Services
P.O. Box 2098
Nicosia

DENMARK
SamfundsLitteratur
Rosenoerns Allé 11
DK-1970 Frederiksberg C

DOMINICAN REPUBLIC
Editora Taller, C. por A.
Restauracion e Isabel la Catolica 309
Apartado Postal 2190
Santo Domingo

EL SALVADOR
Fusades
Edifico La Centro Americana 6o. Piso
Apartado Postal 01-278
San Salvador 011

EGYPT, ARAB REPUBLIC OF
Al Ahram
Al Galaa Street
Cairo

The Middle East Observer
8 Chawarbi Street
Cairo

FINLAND
Akateeminen Kirjakauppa
P.O. Box 128
SF-00101
Helsinki 10

FRANCE
World Bank Publications
66, avenue d'Iéna
75116 Paris

GERMANY, FEDERAL REPUBLIC OF
UNO-Verlag
Poppelsdorfer Allee 55
D-5300 Bonn 1

GREECE
KEME
24, Ippodamou Street Platia Plastiras
Athens-11635

HONG KONG, MACAO
Asia 2000 Ltd.
6 Fl., 146 Prince Edward Road, W.
Kowloon
Hong Kong

HUNGARY
Kultura
P.O. Box 139
1389 Budapest 62

INDIA
Allied Publishers Private Ltd.
751 Mount Road
Madras - 600 002

Branch offices:
15 J.N. Heredia Marg
Ballard Estate
Bombay - 400 038

13/14 Asaf Ali Road
New Delhi - 110 002

17 Chittaranjan Avenue
Calcutta - 700 072

Jayadeva Hostel Building
5th Main Road Gandhinagar
Bangalore - 560 009

3-5-1129 Kachiguda Cross Road
Hyderabad - 500 027

Prarthana Flats, 2nd Floor
Near Thakore Baug, Navrangpura
Ahmedabad - 380 009

Patiala House
16-A Ashok Marg
Lucknow - 226 001

INDONESIA
Pt. Indira Limited
Jl. Sam Ratulangi 37
Jakarta Pusat
P.O. Box 181

IRELAND
TDC Publishers
12 North Frederick Street
Dublin 1

ISRAEL
The Jerusalem Post
The Jerusalem Post Building
P.O. Box 81
Romema, Jerusalem 91000

ITALY
Licosa Commissionaria Sansoni SPA
Via Benedetto Fortini, 120/10
Casella Postale 552
50125 Florence

JAPAN
Eastern Book Service
37-3, Hongo 3-Chome, Bunkyo-ku 113
Tokyo

KENYA
Africa Book Service (E.A.) Ltd.
P.O. Box 45245
Nairobi

KOREA, REPUBLIC OF
Pan Korea Book Corporation
P.O. Box 101, Kwangwhamun
Seoul

KUWAIT
MEMRB
P.O. Box 5465

MALAYSIA
University of Malaya Cooperative
    Bookshop, Limited
P.O. Box 1127, Jalan Pantai Baru
Kuala Lumpur

MEXICO
INFOTEC
Apartado Postal 22-860
14060 Tlalpan, Mexico D.F.

MOROCCO
Societe d'Etudes Marketing Marocaine
12 rue Mozart, Bd. d'Anfa
Casablanca

NETHERLANDS
InOr-Publikaties b.v.
P.O. Box 14
7240 BA Lochem

NEW ZEALAND
Hills Library and Information Service
Private Bag
New Market
Auckland

NIGERIA
University Press Limited
Three Crowns Building Jericho
Private Mail Bag 5095
Ibadan

NORWAY
Narvesen Information Center
Bertrand Narvesens vei 2
P.O. Box 6125 Etterstad
N-0602 Oslo 6

OMAN
MEMRB Information Services
P.O. Box 1613, Seeb Airport
Muscat

PAKISTAN
Mirza Book Agency
65, Shahrah-e-Quaid-e-Azam
P.O. Box No. 729
Lahore 3

PERU
Editorial Desarrollo SA
Apartado 3824
Lima

PHILIPPINES
National Book Store
701 Rizal Avenue
P.O. Box 1934
Metro Manila

POLAND
ORPAN
Patac Kultury i Nauki
00-901 Warszawa

PORTUGAL
Livraria Portugal
Rua Do Carmo 70-74
1200 Lisbon

SAUDI ARABIA, QATAR
Jarir Book Store
P.O. Box 3196
Riyadh 11471

**SINGAPORE, TAIWAN, BURMA,
BRUNEI**
Information Publications
    Private, Ltd.
02-06 1st Fl., Pei-Fu Industrial
    Bldg.
24 New Industrial Road
Singapore 1953

SOUTH AFRICA
For single titles:
Oxford University Press Southern
    Africa
P.O. Box 1141
Cape Town 8000

For subscription orders:
International Subscription Service
P.O. Box 41095
Craighall
Johannesburg 2024

SPAIN
Mundi-Prensa Libros, S.A.
Castello 37
28001 Madrid

SRI LANKA AND THE MALDIVES
Lake House Bookshop
P.O. Box 244
100, Sir Chittampalam A. Gardiner
    Mawatha
Colombo 2

SWEDEN
For single titles
Fritzes Fackboksforetaget
Regeringsgatan 12, Box 16356
S-103 27 Stockholm

For subscription orders
Wennergren-Williams AB
Box 30004
S-104 25 Stockholm

SWITZERLAND
For single titles
Librairie Payot
6, rue Grenus
Case postal 381
CH 1211 Geneva 11

For subscription orders:
Librairie Payot
Service des Abonnements
Case postal 3312
CH 1002 Lausanne

TANZANIA
Oxford University Press
P.O. Box 5299
Dar es Salaam

THAILAND
Central Department Store
306 Silom Road
Bangkok

**TRINIDAD & TOBAGO, ANTIGUA
BARBUDA, BARBADOS,
DOMINICA, GRENADA, GUYANA,
JAMAICA, MONTSERRAT, ST.
KITTS & NEVIS, ST. LUCIA,
ST. VINCENT & GRENADINES**
Systematics Studies Unit
55 Eastern Main Road
Curepe
Trinidad, West Indies

TURKEY
Haset Kitapevi, A.S.
Davutpasa Caddesi
Sergekale Sokak 115
Topkapi
Istanbul

UGANDA
Uganda Bookshop
P.O. Box 7145
Kampala

UNITED ARAB EMIRATES
MEMRB Gulf Co.
P.O. Box 6097
Sharjah

UNITED KINGDOM
Microinfo Ltd.
P.O. Box 3
Alton, Hampshire GU34 2PG
England

URUGUAY
Instituto Nacional del Libro
San Jose 1116
Montevideo

VENEZUELA
Libreria del Este
Aptdo. 60.337
Caracas 1060-A

YUGOSLAVIA
Jugoslovenska Knjiga
YU-11000 Belgrade Trg Republike

ZIMBABWE
Longman Zimbabwe
P.O. Box 5T 125, Southerton